Disrupt It

Disrupt It

How architecture, engineering, and construction executives
can transform their organizations in the age of AI disruption

DR. SAM ZOLFAGHARIAN
DR. MEHDI NOURBAKHSH

GRAMMAR
FACTORY
— ESTᴰ 2013 —

Grammar Factory Publishing
MacMillan Company Limited
25 Telegram Mews, 39th Floor, Suite 3906
Toronto, Ontario, Canada
M5V 3Z1

www.grammarfactory.com

Disrupt It: How Architecture, Engineering, and Construction Executives Can Transform Their Organizations in the Age of AI Disruption / Sam Zolfagharian and Mehdi Nourbakhsh.

Paperback ISBN 978-1-998528-05-9
Hardcover ISBN 978-1-998528-07-3
eBook ISBN 978-1-998528-06-6

1. TEC003070 Technology & Engineering / Artificial Intelligence.
2. TEC005000 Technology & Engineering / Construction / General.
3. BUS090000 Business & Economics / Strategic Planning.

Production Credits
Cover design by Designerbility
Illustrations by Pierre Langlois
Book production and editorial services by Grammar Factory Publishing

Grammar Factory's Carbon Neutral Publishing Commitment
Grammar Factory Publishing is proud to be neutralizing the carbon footprint of all printed copies of its authors' books printed by or ordered directly through Grammar Factory or its affiliated companies through the purchase of Gold Standard-Certified International Offsets.

Disclaimer
The material in this publication is of the nature of general comment only and does not represent professional advice. It is not intended to provide specific guidance for particular circumstances, and it should not be relied on as the basis for any decision to take action or not take action on any matter which it covers. Readers should obtain professional advice where appropriate, before making any such decision. To the maximum extent permitted by law, the author and publisher disclaim all responsibility and liability to any person, arising directly or indirectly from any person taking or not taking action based on the information in this publication.

*To our beloved parents, For your
endless support, love, and encouragement.
You've always believed in us and inspired
us to pursue our dreams. Your guidance
and values have been our greatest gifts.*

With love and gratitude,

SAM AND MEHDI

Contents

PREFACE

The idea for this book was born out of our passion for the architecture, engineering, and construction (AEC) industry and our belief in the transformative power of artificial intelligence (AI). We have seen the challenges and changes in the industry and believe AI has the potential to significantly improve how we work.

We wrote this book out of both frustration and hope. The frustration comes from seeing the gap between AI's potential and how it's actually being used in the AEC industry. Despite AI's transformational potential, many AEC companies view it as a tactical initiative primarily used for achieving only efficiencies and productivity gains.

Our hope is what inspired us to write this book. We believe that AI can do much more than make incremental improvements in operations; it can radically transform the way our industry works—for the better.

Our previous book, *Augment It: How Architecture, Engineering and Construction Leaders Leverage Data and Artificial Intelligence to Build a Sustainable Future*, focused on understanding AI and its potential in the AEC industry. This book goes further and provides a practical framework for leveraging AI's transformational potential, building around three main pillars: culture as the foundation, business model as a value generator, and operating model as an enabler.

Our goal is to move the conversation beyond surface-level uses and explore how AI can be integrated into your company's operations, driving strategic innovation, real value, and growth. We want to see AI being used not just as a tool but as an asset that transforms your company into a strategic innovator.

We hope this book becomes a valuable resource on your journey to drive a transformational shift, challenge the status quo, and disrupt your business and the industry before being disrupted by others. Let this be your guide to staying ahead of the curve in the new age of innovation.

Dr. Samaneh Zolfagharian
Dr. Mehdi Nourbakhsh

INTRODUCTION

The film *Iron Man* (2008) tells the story of Tony Stark, a brilliant inventor and the CEO of weapons manufacturer Stark Industries. In an early scene, Tony is captured by terrorists in Afghanistan, and while he is being held prisoner, he suffers a severe chest injury caused by shrapnel. To save his own life and escape captivity, Tony builds an electromagnet to prevent the shrapnel from reaching his heart.

Unlike many superheroes, Tony does not possess inherent superpowers. His abilities come from his intellect and technological prowess, which allowed him to create the Iron Man suit. The suit made it possible for Tony to do things he could not have done before; it was a new tool that gave him mobility, combat capability, and longevity.

Like Tony's Iron Man suit, we see AI as a powerful new capability for your organization. If this capability is integrated into your organization effectively, it can bring tremendous value and market differentiation. However, if your competitors utilize it first, it could cause a significant disruption to your business.

Going through technology disruptions is not new to the AEC industry. Over the past several decades, we have gone from making drawings with ink and mylar to using computer-aided design (CAD) and building information modeling (BIM). In construction, we have gone from old-fashioned construction to leveraging sensors, drones, and predictive analytics to build on time and within budget. In the case of each of these technological advancements, some companies utilized them and thrived. Others took no action or acted late, and those companies often failed or suffered low-profit margins.

Looking at the technological advancements in the AEC industry, it's clear that the shift from ink and paper to CAD, while significant, didn't fundamentally change the process but did make it digital. The transition to BIM and Cloud, however, drastically changed the processes and the way the industry works. This not only improved efficiency but also brought a new type of coordination and collaboration to the industry, created new jobs such as BIM managers and BIM coordinators, and changed how we design and build.

We are now transitioning to a new era in which AI is bringing enormous market opportunities to every industry. The size of the global AI market was estimated at $196.63 billion in 2023 and is projected to grow at a compound annual growth rate (CAGR) of 36.6% from 2024 to 2030 (1). Because of AI's huge potential to reshape businesses, companies across many industries have accelerated their adoption of AI. According to research data from McKinsey, IBM, Forbes, and Shortlist, 82% of global companies in 2024 either used or explored AI (2). In comparison, only 20% of companies used AI in their operations in 2017 (2).

This adoption rate is quite different in the AEC industry, which is known for its slow pace of change and innovation (3). So, the question is, what might be holding AEC executives back from leveraging AI in their organizations?

Over the past decade, we have worked closely with CEOs and executives from many AEC companies who were eager to begin their AI journey. We have also delivered hundreds of AI keynotes, talks, workshops, webinars, and roundtable discussions specifically focused on AI in the AEC industry. At the start of these sessions, we often ask: "What questions do you have about AI that we should address today?" We've received thousands of questions from AEC executives and, after analyzing them, found that they centered around three key themes: culture, business, and operation. Here's a summary of the most common questions:

Culture

- How does AI impact our people?
- How should we prepare our people?
- What are the cultural initiatives we should take?

Business

- How can we be on the cutting edge, not the bleeding edge?
- How can we prepare our business for AI?
- How can AI help us generate more revenue and market share?

Operation

- How can we utilize our data in the best way?
- How will our teams work together in the age of AI?
- How will AI impact our processes?

These recurring themes, related to culture, business, and operation, made us realize that many AEC leaders are dealing with similar AI concerns. That's why we decided to write this book. We wanted to answer the questions we've heard repeatedly and provide straightforward advice to help executives like you integrate AI into your business successfully. We hope this book will guide you through the challenges, help you navigate the complexities of AI adoption, and make a real difference in your organization.

In this book, we have captured everything you need to navigate your AI journey. We elaborate on how to prepare your culture, business model, and operating model for the age of AI. In doing so, we offer practical advice, real-world examples, and actionable steps. You will find insights from our experiences working with industry leaders, along with answers to the most common questions we've heard. Whether you are just starting out or looking to refine your strategy, this book will guide you in making AI work for your organization. Here is what to expect...

In Part 1, we discuss the six innovation waves that occurred during and after the Industrial Revolution—from textile production in factories to the age of AI. You will learn how successful companies leveraged these technologies to achieve what was previously impossible or unthinkable. We also describe the patterns found in these technological waves and the new opportunities presented in the age of AI. Next, we share various ways that AEC companies respond to AI and how and why you should act now. Companies that do not act risk jeopardizing their market position and competitiveness. This can lead to falling behind in efficiency gains, capacity growth, and meeting evolving client expectations.

In Part 2, you will learn how to lead your company in the age of AI by establishing culture as the foundation, business model as a value creator, and operating model as an enabler. You will learn how to set a budget and the right expectations for the ROI of your AI initiative. You will learn how to equip your people and organization with the right mindset and emotions to deal with the uncertainties of AI. Next, you will learn the impacts of AI on your business model through the lens of tactical and transformational business opportunities, where you can create new services for your clients, expand your market segment, or create new revenue models. Finally, you will learn how to change your operating procedures and workflows to create consistency in your data, which is your organization's most critical asset in the age of AI. You will also learn about the network effects of AI opportunities and why they matter.

But that's not it! Throughout this book, you'll hear from several visionary CEOs and executives about how they are preparing their companies for this new era. You will hear from Vince DiPofi, the CEO of SSOE, about how he's been creating a culture of innovation in his company; from Andy McCune, the CEO of Wade Trim, on how they leverage AI to get new market share; and from Josh McDowell and Matt

Butts, principals and board members at Mackenzie, on the importance of creating new operational workflows to get the most out of their digital assets—data.

Who is this book for?

Returning to San Francisco on a flight one day, we were reflecting on a discussion we'd had with more than fifty technology leaders and IT professionals from the largest AEC firms in the U.S. Our takeaway was that most of them were asking the same question: "How do we get buy-in from our CEO, senior executives, and the board when it comes to our AI initiatives?"

They were very frustrated by one of the common misconceptions in the AEC industry: that the technology department should manage and lead AI initiatives. Their CEOs and executives did not recognize that more than half of AI initiatives fail because people and processes within organizations are not ready. This underscores that many AI-related challenges in our industry are not technical. They are people and process issues. That's why, contrary to common belief, AI initiatives should be driven by the CEO and executed by operations teams, with strong support from technology teams. If your organization does not have a CEO role, a principal executive appointed by the board can lead this effort. We realized that *Augment It*, written as a primer AI book for AEC executives—specifically technical professionals such as IT executives, CIOs, and CTOs—did not address the importance of why CEOs, board members, and non-technical leaders should lean in.

Given the importance of aligning AI initiatives with your organization's culture, business models, and operations and the need for strong leadership to drive these efforts, we wrote this book primarily for CEOs, principals, and executives of AEC companies. In addition, board members, non-technical executives (heads of finance, marketing, HR, heads of business units and operations, etc.), and leaders who make critical business decisions for their companies can also benefit from this book.

Throughout this book, we use the terms "CEOs" and "executives" to refer to the primary leaders of AEC organizations. However, we understand that not all AEC companies have traditional hierarchical structures. In organizations with flat structures—where leadership is shared among board members, principals, or heads of business units—it's essential to appoint a dedicated senior leader to oversee AI initiatives. Please interpret "CEO" and "executives" in a way that aligns with your organization's structure.

Who is this book NOT for?

In this book, we included examples and anecdotes from various industries like manufacturing, technology, and farming, but the primary focus is the AEC industry. We wrote this book for AEC executives looking to grow their established companies in the age of AI. That said, even if you're not an AEC executive, you may still find valuable perspectives applicable to your industry.

If you are a student, academic, intern, researcher, or not in a leadership and executive role, you might not find this content especially useful. Moreover, if you are a current or future startup founder or an entrepreneur, this book may not be directly relevant, as the advice given is tailored for executives in established businesses. However, you may still find value in the strategies and insights shared in this book that can inspire growth approaches for ventures at any stage.

What is this book NOT about?

This book does not cover the technical aspects of AI, such as machine learning, deep neural networks, generative AI, or other branches of AI. It also does not cover any AI applications or use cases in the AEC industry. This book presumes that readers already have a basic understanding of AI, so you will not find definitions or explanations of what AI is or how it works. If you want to learn how AI algorithms work, you will need a different book because this one does not cover that. Instead, our focus is on leveraging strategic innovation to transform your organizations for the age of AI, leaving aside the technical. *Augment It* might be a good starting point if you want to understand AI and its applications in AEC and know how to set an AI strategy for your company.

Additionally, this book does not cover AI risks and governance. For more on these topics, refer to the reading list provided in our companion workbook, available at www.disruptit-book.com.

Our story

Our careers so far have been very similar, yet we have followed distinct paths. In the first part of our careers, we both worked as structural engineers designing buildings and as construction managers. Our professional experience gave us a deep understanding of the challenges and opportunities within the AEC industry. Having practical knowledge, we started the second part of our careers by pursuing higher education. We obtained master's degrees in construction management and our Ph.D.s in building construction and design computing from the Georgia Institute of Technology (Georgia Tech). We were passionate about the R&D aspects of higher education, where we could experiment with various technological tools both inside and outside of the design and construction space, including virtual reality, augmented reality, drones, 3D printing, and many more.

While working on our Ph.D.s, Mehdi enrolled in a dual-degree program to earn his master's in computer science at Georgia Tech and learn more about the fascinating world of AI. Meanwhile, Sam focused on processes and frameworks for data exchange between various BIM applications, a project funded by the national BIM standard, several other associations, and software companies.

In the third phase of our careers, we both joined Autodesk. Sam was a product manager, developing technical solutions for industrialized construction and later for Autodesk's cloud platform. She led a team of software engineers leveraging AI in developing manufacturing design tools and later applying AI to personalize users' experiences across Autodesk products. Mehdi worked as a research scientist and was later promoted to innovation manager, leading the development of the next generation of AI applications in the AEC and manufacturing industries. He published eight U.S. patents related to AI applications in these fields.

Throughout our journey, we have shared a similar passion for innovation and pushing boundaries. We have always been fascinated by the potential of technology to transform human lives and make the world a better place, not only for us but for generations to come. Our dreams have been fueled by a commitment to lifelong learning and a vision of a future where technologies and innovation transform the AEC industry into a more sustainable one. Throughout our journey, we have faced many challenges and had our own ups and downs, but our dedication to our belief in the power of innovation and leaving a legacy has kept us moving forward.

After fruitful years at Autodesk, we entered the fourth phase of our careers by starting YegaTech, an innovation consulting company. Our mission at YegaTech is to disseminate knowledge and help AEC companies navigate the world of AI and innovation. We assist them in creating the right AI strategies, governance, and implementation plans to help their companies thrive while also preparing their organizations for the future of work through forward-thinking practices and adaptability.

Working with many CEOs and AEC executives, we have experienced firsthand the importance of having a guide in a company's AI journey. We have seen AI projects fail mainly because AEC executives could not fully grasp how to lead AI initiatives, as they viewed AI as "just another technology initiative" rather than a "transformative change."

We have also observed how other executives, with the right guidance, transformed their companies into innovators. By creating a culture of innovation, they were able to change their employees' mindset from being fearful and skeptical about AI to being excited and energized by its potential. Those employees then brought many innovative ideas to their businesses and pushed the executive to invest more in technology and support transformative initiatives. We saw companies flourishing by taking the right steps to disrupt themselves, challenging the status quo, and stepping out of their comfort zones.

In helping these companies in their transformational journey, we realized there is a huge opportunity to transform our industry by helping companies develop their own "Iron Man suit." That's why we wrote this book. This book will help you flawlessly navigate the current age of AI disruption and become the next industry leader. This book is about making AI work for you.

Before we go any further, let's bust a common myth. Many AEC executives believe that AI is only for large companies. However, with recent advancements, we believe the playing field has leveled. In fact, smaller companies often have a significant advantage due to their size and agility, which allows them to adopt and innovate with AI more flexibly than larger organizations. Whether leading a small team or a large organization, we aim to provide you with a practical framework and the confidence to successfully operationalize AI in your company. We hope this book becomes a valuable resource on your journey to innovation and growth in the age of AI disruption.

Since the world of AI is changing so fast, we created a companion workbook containing case studies, reading recommendations, and additional resources to help you execute the ideas shared in this book. You can access the companion workbook here: www.disruptit-book.com.

Are you ready to start?

PART 1

Navigating AI Disruption

How did we end up here as an industry? Just a few centuries ago, we were in the master builder era, and now we are an industry with thousands of broken pieces. While the top three percent of the industry is innovating at an unprecedented rate, the rest is lagging and ignoring innovation.

For those who innovate, disruption is no longer an occasional act but a constant force driving their progress and reshaping their business. To grow and stay ahead of the curve, these companies are trying to understand the new wave of innovation (the age of AI) and leverage it to potentially transform how their businesses operate.

This section of the book explores the concept of disruption and strategic innovations in the context of AI. We begin by discussing the importance of disruption as a driver of progress, tracing its evolution through waves of innovation and identifying patterns. We discuss the disruptive forces of the latest wave (AI) and how they impact all industries. We then narrow our scope to AEC and explore potential opportunities and transformational shifts AI brings to this industry.

Next, we discuss how to navigate the sixth wave of innovation. We also explain why your response should be to rethink your business models and operations and do more than just adopt new AI tools. Doing so can create lasting value and help you stay ahead in a complex and dynamic market competition.

CHAPTER 1

The Necessity of Disruption

When Circle K, a North American chain of convenience stores, filed for bankruptcy in May 1990, the board brought in John to streamline the company's operation. John was known for his vision and power to execute. Under his leadership, he privatized the company, closed the unprofitable stores, and focused on the profitable ones. In just a few years, he transformed Circle K into a successful company and sold it to Tosco for $710 million. Because of John's leadership and execution, the creditors received 100% of the dollars they invested, and the equity group investors earned four times their investment in under three years.

Circle K was not the only business that John turned around; he's done the same for several other big brands, like 7-Eleven and Taco Bell. After multiple accomplishments, he took on a new challenge by joining a struggling brick-and-mortar business in the entertainment industry, which had been in rapid decline under its former leadership. Soon after he joined as the CEO, John launched a successful IPO and raised $465 million in cash to scale his operations and open more stores.

This was back in the early 2000s, and while John was busy helping the company grow, others, especially startups, were beginning to leverage a new disruptive technology called the Internet. At that time, the

Internet's full potential was not understood, and no one really knew how it could fundamentally impact brick-and-mortar businesses. Also, since Internet adoption was still in its early stages, many companies saw the online business model as risky because it required massive capital investment and produced low initial returns.

One of John's competitors was a startup in Scotts Valley, California. The startup wanted to offer the same services online that John's company offered in stores. However, the company was struggling to get traction and grow as quickly as they had hoped. They had invested everything in building an online business but did not have enough customers, and at the rate they were acquiring new customers, they were on track to lose 50 million dollars by the end of the year 2000. The co-founders (Marc and Reed) and the CFO (Barry) had no choice but to meet with John and try to convince him to acquire them.

One day in September 2000, Marc, Reed, and Barry met with John and his team. During the meeting, Reed proposed that the companies join forces.

"We will run the online part of the combined business. You will focus on the stores. We will find the synergies that come from the combination, and it will truly be a case of the whole being greater than the sum of its parts," Reed said.

After John heard Marc, Reed, and Barry's proposal for a merger, his immediate response was, "The dot-com hysteria is completely overblown." His general counsel, Ed, added that online business models were unsustainable and never made money. After some back-and-forth and debate, Ed asked, "If we were to buy you, what were you thinking? I mean, a number."

"Fifty million," Reed said.

The meeting went downhill quickly after that. Marc, Reed, and Barry returned home empty-handed, thinking they had lost their chance to be acquired and wondering how they could survive the business and revenue challenges they were facing.

The story you have just read is from *That Will Never Work, the birth of Netflix and the amazing life of an idea* by Netflix co-founder Marc Randolph (4). It describes how Marc, along with Netflix's other co-founder, Reed Hastings, and its then-CFO, Barry McCarthy, met with John Antioco at Blockbuster headquarters in Dallas.

Even though John Antioco had an impressive record in turning big businesses around, he ignored technological disruptors like the Internet and streaming services. In just a few years, Blockbuster tried to catch up by investing $200 million to launch Blockbuster online and removing the late fee (which resulted in a $200 million revenue loss), but it was too late. The rest is history.

The lesson learned from this story is:

Leaders who stop innovating and disrupting their businesses jeopardize their companies and shareholders.

Many years later, in 2022, John Antioco wrote, "As I reflect on all of this, I am reminded of Darwin's theory of evolution: 'It is not the strongest of the species that survive, nor the most intelligent, but the ones who are most responsive to change.'" Adaptability is the key to survival, not just for animals and humans but for businesses as well.

The businesses most responsive to change are the ones that thrive and prosper.

The Netflix and Blockbuster stories are examples of companies that react differently to disruptive innovations, such as the Internet and streaming. Throughout history, we've seen many disruptive innovations, such as the steam engine, railroad, electricity, automobiles, and the Internet, to name a few. Each innovation was introduced in an era known as the "innovation wave." In this context, an innovation wave is defined as the process of technological and economic development characterized by a period of rapid innovation followed by a period of slower growth and consolidation. Each innovation wave typically involves the introduction of new technology, its widespread adoption, and the eventual maturation of both the technology and the associated industries.

Typically, in each innovation wave, many new businesses are created, and many companies thrive, while many fall behind because they fail to adapt quickly.

Those companies that do not change and disrupt themselves often face disruption from more agile competitors.

In this chapter we'll discuss six innovation waves and their patterns over time, then deep dive into the sixth innovation wave: the age of AI. We will provide a brief history of AI and the significant developments that got us here.

But before discussing multiple waves of innovation, let's establish the definitions of "innovation" and "disruption" as they are used in this book.

Innovation refers to the introduction of new ideas, products, services, or processes that create value or improve existing solutions.

Disruption occurs when an innovation significantly alters or replaces existing businesses, markets, and industries. It can happen in a controlled way in your business or be caused by an external entrant. New entrants often challenge established players by offering products or services that are cheaper, more convenient, or better suited to current client needs. For example, ride-sharing apps like Uber disrupted the traditional taxi industry by providing more accessible and user-friendly experiences, and e-commerce platforms like Amazon disrupted the retail industry by offering a convenient and often more affordable shopping experience.

Innovation is the driving force that can lead to disruption when it significantly changes how businesses and industries operate. Disruption is the process through which innovations reshape businesses and markets, often resulting in the decline of traditional businesses and the rise of new market leaders.

Now, let's explore various innovation waves in history and discuss their patterns over time.

The evolution of disruption: Six waves of innovation

The history of technological innovations (illustrated in Figure 1.1) has been full of patterns, reminding us of Mark Twain's famous quote: "History doesn't repeat itself, but it does rhyme." As we explain the six waves of innovation that occurred during and after the Industrial Revolution, we encourage you to look for recurring patterns and see how they evolve over time.

Before the eighteenth century, economies were primarily agrarian, and manual labor and animals were the primary production sources. During the Industrial Revolution, one of the early waves of innovation (approx. 1785 –1845) was the shift from hand production to machinery and the rise of factory systems. This shift led many people to migrate from rural areas to urban centers in search of factory jobs, speeding up the development of industrial societies, which was further expedited by James Watt's steam engine invention.

The second wave of innovation (approx. 1845—1900) was about advancements in transportation and communication. The expansion of railways and steam-powered ships revolutionized trade and economic integration. Samuel Morse's telegraph enabled rapid long-distance communication, leading to global trade growth and accelerated economic development.

The third wave of innovation (approx. 1900—1950) was about widespread electrification, which revolutionized industries and daily life through electric machinery, lighting, and appliances. The rise of the automobile industry reshaped transportation, and Henry Ford's mass production techniques made consumer goods accessible. The first flight of an engine-powered aircraft, a feat achieved by the Wright brothers in 1903, marked the start of aviation and set the stage for further industrialization, urbanization, and economic growth.

The fourth wave of innovation (approx. 1950 –1990) marked significant advancements in petrochemicals, electronics, and computing. This period was characterized by the production of synthetic materials and the widespread use of fossil fuels, which impacted economic growth but raised environmental concerns. The use of the transistor during this wave revolutionized electronic devices and consumer electronics. In the latter half of this wave, the rise of personal computers and the semiconductor industry laid the foundation for the digital revolution. They set the stage for the next wave, the Information Age.

The fifth wave of innovation (approx. 1990 –2020) was defined by information technology and the widespread adoption of the Internet. During this period, advances in telecommunications, software development, and digital communication reshaped industries and shifted how we generate, process, store, and share information. Companies had to adapt to online systems and digital marketing and change their business models to more agile and technology-driven systems. This wave's focus on digitalization and connectivity has created a globally connected digital society and transformed everyday business processes and lifestyles.

The sixth wave of innovation (approx. 2020—present) is characterized by the AI revolution and the mass availability of AI agents designed to improve productivity and efficiency across industries. According to PwC, the total economic impact of AI on the global economy will reach $15.7 trillion by 2030 (5). The International Monetary Fund (IMF) reports that:

In advanced economies, about sixty percent of jobs may be impacted by AI. Roughly half the exposed jobs may benefit from AI integration, enhancing productivity. For the other half, AI applications may execute key tasks currently performed by humans, which could lower labor demand, leading to lower wages and reduced hiring (6).

Given the recent and relatively unknown nature of this wave of innovation, particularly within the AEC industry, we will now address a pressing question many AEC executives ask: "How did we get here?"

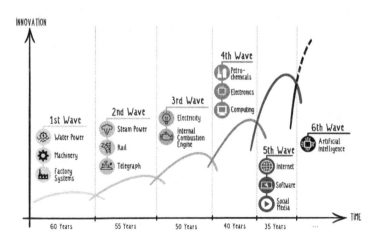

Figure 1.1: Six innovation waves, adapted from Edelson Institute (7)

AI: the sixth wave of innovation

In our discussions with AEC professionals, we often hear the question: "How did AI come out of the blue?" The short answer is, "It did not." AI has been around since the 1950s, but it's only in recent years that it has gained widespread attention. The story of AI development is a fascinating one, rooted in decades of research and innovation.

It all started with one simple test to answer one important question—a question that many executives still ask today: "How can you say this *is* AI, or is *not* AI?" This was a driving question of British mathematician Alan Turing's research in 1950. He developed a method for testing computer systems that mimicked human intelligence (8). In this test, known as the Turing Test, a human judge uses natural language conversation and question answering with one human and one machine, without seeing them. If the judge cannot distinguish the machine from the human consistently and reliably, the machine passes the test. This test became foundational in examining artificial intelligence systems.

A year later, in efforts to create computers that could emulate human behavior, Marvin Minsky and Dean Edmonds developed the first artificial neural network, SNARC (Stochastic Neural Analog Reinforcement Computer), which simulated how rats find their way through a maze (9). Their work was inspired by Warren S. McCulloch's and Walter Pitts' research on the abstract model of the brain and how neurons work together to enable us to learn (10). This work laid a foundation for the field of AI, demonstrating computer systems that could mimic the human learning process.

In the 1950s and 1960s, the field of AI showed lots of promise and attracted noticeable funding from governments and research institutes. Despite considerable theoretical progress in AI, practical applications were limited or sometimes even impossible. Eventually, funding agencies withdrew their support and paused funding for AI projects, leading to the first AI 'winter' between the mid-1970s and mid-1980s.

Following the first AI winter in the 1980s, one of the major AI advancements was the development of expert systems, which allowed computer systems to emulate human expert knowledge. However, these systems' popularity declined due to scalability issues and maintenance difficulties. Again, this led to a roadblock in AI advancement, which caused the second AI winter from the late 1980s to the mid-1990s.

To understand why early AI efforts did not succeed, we need to understand the three main components empowering AI systems development: data, computing power, and algorithms (math). Generally speaking, you can build a better AI system by utilizing larger data sets, better algorithms, and more computing power. Let's look back and see how each of these ingredients has changed over time.

Data

Let's look back to the late 1990s, when Internet usage expanded rapidly. As more people went online, they generated more data by interacting on websites, especially through smartphones and social media platforms. The increase in online presence and interactions generated ever larger volumes of data and led to the coining of the term 'big data' in the early 2000s. And this online activity continues to grow.

Computing power

As the Internet continued to expand and generate more and more data in the early 2000s, the gaming industry became a big driver in pushing the boundaries of hardware, as gamers consistently look for the best and latest games with higher-quality graphics. As a result, companies like NVIDIA focused on developing advanced graphics processing units (GPUs), which not only revolutionized the gaming and graphics industries but also impacted AI research and development through access to higher computational power. This progress in chip technologies has enabled the development of more complex algorithms and faster processing during AI training and operation.

Algorithms

As more data and computing power have become available, AI researchers and scientists have started improving the algorithms and mathematics behind the creation of AI systems. Some of the most frequently used types of AI algorithms are known as neural networks, which are loosely based on how our brain neurons work (in a continuation of Marvin Minsky's research discussed earlier). Neural networks typically consist of an input layer, multiple hidden layers, and an output layer (see Figure 1.2). Each layer transforms the input data to identify patterns and features, with deeper layers typically capturing more complex aspects of the data. However, in the past, due to limited computational power and data availability, the number of layers was typically fewer than four or five. So, the accuracy and capabilities of AI models were often low.

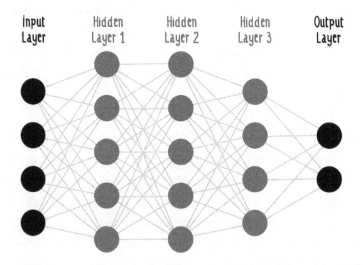

Figure 1.2: The architecture of a neural network

However, the invention of Generative Adversarial Networks (GANs) in 2014 was a game-changer in developing generative AI systems that could generate images (11). Many tech companies started to take advantage of this technology, including the founders of OpenAI, who saw an opportunity to create an AI system using the massive amount of data available on the Internet. They hired some of the brightest minds in AI and secured billions of dollars to train their AI system.

In 2020, OpenAI released GPT3, a large language model (LLM) capable of answering questions and translating text between languages. The following year, they released DALL-E, an AI system capable of generating photo-realistic images based on a textual description (prompt). In 2022, they released ChatGPT, a chatbot for answering questions, summarizing text, translating language, writing computer programs, and more. ChatGPT quickly became the fastest-growing consumer application in history. It had over a million users in the first five days of release and 100 million active users two months after its release. To put this in perspective, it took TikTok nine months and Instagram two and a half years to reach 100 million users (12).

The difference between these large language models and other deep learning models of the 2010s is that these models can be used as a foundation for building a wide array of technology solutions. For instance, these foundational models can be customized to answer customer support questions about technical troubleshooting, medicine, climate change, and more. Previously, different AI models had to be trained for each specific application, but now, a single foundational model can be adapted across various applications.

These foundational models, including large-scale neural networks, are possible because more data, higher computational power, and more advanced algorithms are now available. Foundational models, like GPTs, are trained on a massive amount of data and neural networks. As a result, these foundational models can generate high-quality human-like text, images, and videos in a short time. With all these advancements in AI, many people believe that we are at the beginning of an AI spring—a period in which more funding becomes available from governments, funding agencies, and large companies with budgets of billions of dollars. In other words, this is just the beginning of many technological changes in many industries, including the AEC industry and, therefore, your organization.

Thriving through disruptions

Let's step back to look at all six waves of innovation. While many companies have been created and failed over the past two centuries, relatively few have been able to adapt to change and remain in business for a long time. Let's review John Deere as a perfect example of how a business can adapt and thrive by leveraging opportunities created in each wave of innovation (See Figure 1.3 for illustration):

John Deere was founded in 1837 with the invention of the self-scouring steel plow in Illinois. During the Industrial Revolution, the company pivoted to more complex agricultural machinery, including tractors and combines, which boosted farming productivity. Later, with the invention of the Internet and digitalization, John Deere integrated GPS, satellite guidance, and telematics into its machinery, enabling farmers to optimize planting and harvesting and better manage their fields. With the recent wave of innovation and AI, John Deere took its business to the next level by integrating advanced computer vision into its equipment. This led to the AI-powered See & Spray product, which increases precision in identifying and targeting weeds, minimizing environmental impact and using resources better. In response to the growing global population and labor shortage, the company invested in its next strategic innovation, fully autonomous farming, which it hopes to achieve by 2030.

John Deere has continually evolved from its first innovation, the steel plow, to the adoption of industrial machinery, digital farming, and now AI-driven agriculture. By adopting disruptive innovation, the company has remained at the forefront of agricultural innovation for nearly two centuries!

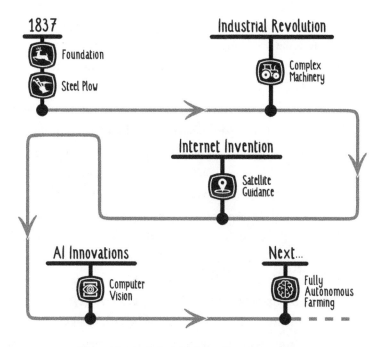

Figure 1.3: John Deere's business disruption over time

Take a moment to reflect on the six waves of innovations discussed so far. Have you noticed any recurring trends or patterns?

Decoding the disruption patterns

By taking a closer look at the trends evident throughout these waves of innovation, three primary patterns emerge that require critical consideration:

1. **From bits to bytes.** Most early innovation waves had physical components (railroads, cars, airplanes, etc.) that required raw material extraction and processing. Typically, components were

developed in original equipment manufacturer (OEM) facto-ries and assembled in other manufacturing facilities. During the fourth wave of innovation in 1955, companies like General Motors, Exxon Mobil, United States Steel Company (the first billion-dollar company in the stock market), and General Elec-tric dominated the industrial landscape and economic growth. What do they have in common? They were heavily invested in the process of transforming bits into physical components. However, these companies have not maintained their presence on the list of top industry leaders because the last two waves of innovation have been about software technology, data, and AI. In these waves, tech companies like Apple, Amazon, Microsoft, NVIDIA, and Alphabet led and continue to lead the stock market.

2. **From slow pace to fast.** The inventions during the first couple of innovation waves used to take five or six decades to become scalable and be adopted at a mass level. This was due to several factors, including limited infrastructure. However, as we moved toward the fifth and sixth waves, the length of each wave was reduced from five or six decades to just a couple. And the trend shows that future innovation waves will be even shorter!

   This accelerated pace of innovation is both a challenge and an opportunity for business leaders. While it can lead to the disruption of existing business models and industries, it also presents opportunities for growth, transformation, and com-petitive advantage.

3. **From chaos to order.** Earlier waves of innovation brought order to the chaos of the physical world. Innovations like industrial textile production, railroad transportation, and Henry Ford's assembly line automated many repetitive and time-consuming tasks and created new opportunities. As we have transitioned from the physical world into the digital world over the past several decades, we have created a lot of repetitive and time-consuming

digital tasks that employees need to perform daily. Just as early innovations brought some order to the physical world, AI can bring order to the chaotic digital world we have created. It can be leveraged to automate complex and repetitive tasks and improve the metaphorical assembly lines in your company.

So, in summary, we have transitioned from a slow-paced physical world to a fast-paced digital world, where companies that are leveraging the sixth innovation wave (AI) are leading the market. Why is knowing this important for you as a business leader?

According to Joseph Schumpeter, an Austrian-American economist who coined the term "Creative Destruction," each wave of innovation led to the demise and disruption of older industries and economic structures, paving the way for new industries and economic growth. In other words, business leaders who failed to recognize and adapt to technological disruptors ultimately led their companies to irrelevance. Kodak missed the digital photography revolution, and John Antioco, the former CEO of Blockbuster, underestimated the rise of the Internet and streaming services. By the time these companies understood the transformative potential of these new technologies, it was too late to pivot. So, they went out of the game.

But which companies have thrived? Was it the innovators who commercialized their innovations or the companies who figured out how to effectively leverage those innovations? History shows that while we have had some rich innovators, like Henry Ford, a lot of the innovators, such as the Wright brothers or Timothy John Berners-Lee (best known as the inventor of the World Wide Web, the HTML markup language, the URL system, and HTTP) did not get the maximum economic benefits out of their innovation. So, who did?

Most of the economic value of new technology went to companies with visionary leaders who leveraged it to create new or improve existing products or services or reach new markets.

For example, in the early 1900s, icemakers had huge factories making giant blocks of ice that were sold to consumers. In those days, most U.S. households kept their perishable items in iceboxes made out of wood and cooled by these giant blocks of ice. But people wanted something better. In 1927, General Electric (GE) developed the first household fridge, the Monitor-Top refrigerator, changing electric refrigeration from a luxury to an affordable appliance in the U.S. (13).

With over a million refrigerators sold, who do you think got the most economic benefit from this innovation? GE or some other companies?

GE gained a lot of revenue, but the real winners were beverage companies like Coca-Cola. They used this technology to offer a unique experience to their customers by keeping their drinks cold and refreshing.

AI brings us to a similar moment in the most recent wave of innovation. Just as refrigerators changed the beverage industry, AI now enables companies to offer new services, enhance their existing services, and deliver better customer experiences.

The point is that you don't need to become an AI technology provider company, like OpenAI, to maximize the economic benefit of this new wave of innovation. You just need to figure out how this innovation can help your company to differentiate itself in the market. Imagine being a farmer when the railroad was first introduced. An ordinary farmer might have seen the railroads and thought they had nothing to do with farming. A visionary farmer might have seen the same thing but thought about how the new technology could help them distribute their products more efficiently to existing markets or expand their reach into new ones that were previously inaccessible. Those with the latter view would gain unprecedented opportunities to grow and expand their business beyond what was possible before.

Companies that learn how to strategically leverage new technology to go beyond what's possible today are more likely to thrive in a dynamic and ever-evolving business landscape. Companies that see the technology but take no action will be disrupted or struggle to survive.

CONCLUSION

Let's reflect on what we've discussed so far. At the beginning of this chapter, we reviewed the six waves of innovation. Each wave typically involved the introduction of a transformative innovation (e.g. railroad or cloud computing), followed by widespread adoption and the maturation of new technologies and industries. In each wave, companies that leveraged the new technology thrived, while those that ignored it or acted too late struggled or failed.

Unlike textiles, railroads, cars, airplanes, and many other tangible and easily observed advancements, the current wave of innovation, driven by AI, is intangible and less visible, making it harder for us to recognize opportunities. History shows that companies that find ways to leverage new technology receive the most economic value out of the invention.

CHAPTER 2

Impacts of AI on Your Business

One day, a farmer found an eagle's egg on the ground and placed it in his hen's nest. When the egg hatched, a baby eagle was born among the chicks and started living just like them. Every day, the young eagle scratched the ground, pecked for food, and hopped around, never thinking about flying.

One afternoon, a full-grown eagle flew by and noticed this young eagle pecking at the ground. The eagle was surprised and asked, "Why are you pecking the ground when you could be flying high in the sky?"

The young eagle looked up, confused. He had no idea he could do anything else. But with a little encouragement, he spread his wings, moving them up and down until he realized he could lift himself off the ground. Slowly, he started to understand his true potential and left his old life behind, ready to soar.

We share this fable at the beginning of this chapter as a reminder that your company has huge, untapped potential. Like the eagle living on the farm, companies may perform their daily operations unaware of their ability to innovate with the power of AI. That power is like the young eagle's wings, something these companies already possess but don't know can help them to soar.

At this stage, you might not fully understand the new capabilities AI can bring to your organization, but that will change after reading this chapter. Here, we explore the forces of AI and how they can transform your business into a leading industry innovator that delivers tremendous value to your clients if you recognize AI's power.

Disruptive forces of AI advancements

Over the past two decades, we have seen massive capital investment in AI from the public and private sectors. Venture Capitals (VCs) are investing heavily in AI advancement; for example, VCs in the United States have invested over $290 billion between 2019 and 2024, and VCs in China have invested over $120 billion in their country-wide strategic AI plan called "Made in China 2025" (14). With these investments, technology companies are now at the forefront of driving AI innovation, often surpassing academic institutions and universities. While many universities still significantly contribute to AI research, technology corporations have an edge due to their access to data, massive capital for developing large AI models, and strong incentives to gain market share. This will expedite advances in AI technologies, setting new expectations and allowing AI systems to take on more complex tasks. For instance, Microsoft invested billions of dollars in OpenAI, primarily by providing computing power for training AI models in 2023 (15), and will potentially gain $100 billion in the coming years (16). These investments and advancements are part of Microsoft's and OpenAI's global race to dominate this transformative technology that eventually will force businesses to adopt AI or risk being outpaced. These investments will result in two major forces: 1) accelerating the pace of disruption and 2) decreasing the cost of software and AI development. Let's discuss each of these forces.

1. **Accelerating the pace of disruption.** With more data from people and devices connected to the Internet, advancements in computing, and improvements in algorithms, we will see more AI innovation and disruption at a faster pace. Unlike in the past, when every major technological revolution took a decade or more (e.g. the time between the invention of modern steel and using it in modern railroads was several decades), we are now experiencing a continuous wave of market disruption. Tech giants like Microsoft have optimized their operating models to minimize the time between invention and innovation (making the technology available at scale). For instance, Microsoft launched Copilot less than a year after the release of GPT4 by OpenAI. So, we should expect more disruption in the market as the time between inventing new technology and commercialization continues to decrease.

2. **Decreasing the cost of software and AI development.** Over the past couple of decades, technological advancements, availability of reusable code libraries and frameworks, access to affordable cloud computing, and a global talent pool have significantly decreased the cost of software and AI development. For instance, in 2016, developing an AI solution to generate building mass based on a company's historical design cost us about half a million dollars and required a team of six AI engineers and scientists working for a year. In contrast, in 2022, we built an AI solution to detect windows in building facades for roughly $50,000 in about three months with just two AI engineers. Later that year, we developed an AI system to detect cracks in cement for around $5,000 with just one software engineer completing the project in a week. In this project, the software engineer used existing AI models and only had to train the model to detect cracks in concrete and build the user interface. Before OpenAI released the chat with PDF feature on ChatGPT in 2023, one of

our engineers developed a pilot AI system to chat with PDF files that cost roughly $1,200 and was completed in less than a week.

The point is that developing AI and software solutions has become cheaper, easier, and more accessible. Smaller companies and startups can now compete in a field once dominated by large tech companies. Moreover, AI systems can now write software code, which empowers your employees to develop AI solutions to automate repetitive tasks, opening doors to technological innovation in your company.

With billions of dollars in investments in AI innovation, a significant decrease in the cost of software development, and the minimization of the time between invention and innovation, we should expect more AI disruptions in the coming years. But how might these disruptions impact your business?

From passive adoption to active creation

We went to an industry conference where the keynote speaker asked the audience, "How many of you tried to develop technology solutions in the 2010s and failed?" Almost half of the people in the room raised their hands. Then the speaker continued, "You are not a software company; you are a design or construction company. You should not build a software solution."

We agree that almost 99.99% of AEC companies are not traditional tech companies or software companies in the sense that they do not develop software, sell licenses, or maintain software for their clients. Selling tech solutions would require technical teams, support teams, ongoing maintenance, and many other complexities that may not align with your core business. In this context, you are not a software company and likely won't become one.

That said, we define a technology company as any organization that leverages data and AI technology to innovate and enhance its products, services, or operations. In other words, if you have data and access to AI technology, you are a tech company. We firmly believe that every AEC company has the capacity to become a technology company, and those who ignore this opportunity to become one will be irrelevant in the future.

Here's why: with the shortage of professionals and labor in the AEC industry, the old adage "to scale, you need to add more bodies" no longer applies. So, who will lead the future of the AEC industry? It will be companies that, instead of scaling their operations by adding more people, focus on scaling the value of their existing workforce by leveraging the data they produce. These companies treat their data as an asset and utilize AI to augment their organizations' capabilities, providing significantly better services to clients.

The future leaders of the AEC industry will scale not by headcount, but by amplifying the value of their people through data and AI.

Recent AI advancements have created opportunities for all AEC companies, especially smaller ones, to leverage data and scale operations by developing their own technology solutions. How does this work? Let us explain.

For many years, AEC companies have been adopting technology solutions provided by technology companies and startups in the AEC industry, commonly known as AEC tech providers. These AEC tech providers often used to (and still do) utilize hardware and software technologies provided by big tech companies such as Amazon, Microsoft, or Google to develop and offer their services. In other words, they were the intermediary between you and the big tech companies. However, this pattern is changing. Now you have direct access to the latest pre-trained AI models from big tech companies without any intermediary.

Now every company in the AEC industry can develop in-house technology solutions.

As discussed earlier in this chapter, software development and maintenance costs have significantly decreased. Perhaps there was a time when lighting up a Christmas tree was a major electrical project. Back then, you had to hire one or more electricians just to set it up. However, as technology in electronics and manufacturing has advanced, this can now be done quickly and safely at home without needing experts. It has become more like a plug-and-play operation. Similarly, leveraging your data and developing AI solutions is now more like a plug-and-play operation, especially with the presence of no-code app-building tools or pre-built AI models that companies like Microsoft and many startups have made available. The process could be as simple as providing data, calling a few available APIs within your cloud provider's infrastructure, customizing your AI solution, and deploying and running it. This may seem like an oversimplification of how to develop AI solutions, but the point is that this is the direction we are heading.

Nora Swanson, the Director of Knowledge Management & Innovation at JB&B, a US-based MEP (mechanical, electrical, and plumbing) and consulting engineering firm founded in 1915, explains it this way:

The app-making tools developed by companies, like Microsoft, are incredibly user-friendly and easily accessible for every AEC company looking to build its own technology applications. Even our non-developer staff members are able to learn how to create applications simply by watching YouTube tutorials. Many of these applications are now widely used throughout our company. This shows that with the right resources and encouragement, almost anyone can develop useful tools. We've seen firsthand that providing the space for innovation and citizen development can lead to impressive and impactful results.

With the cost of software and hardware decreasing and the capabilities of pre-built AI models increasing, many AEC companies have a huge opportunity to experiment and develop AI solutions tailored to their business needs.

Besides using available plug-and-play tools and APIs, you can write your own code and solutions. You might be thinking, "But we don't have any software developers in our company." Or "We have some developers, but they have no time." However, one important consideration with the recent AI advancements is that English has become the new computer language. If you can communicate in plain English (or any other language of your choice), AI systems can generate the code for you. This is yet another reason why you can easily become a technology company.

English is a new computer language, so every person in your company can write computer codes and is a technologist.

Because code writing is being democratized and commoditized by AI systems, the future belongs to AEC companies with deep subject matter expertise who leverage this technology to develop tech solutions, not just software companies that can write code but lack domain-specific knowledge and expertise. Jensen Huang, the CEO of NVIDIA, emphasized this point at the 2024 World Government Summit, stating:

> ... *for the very first time, you can imagine everybody in your company being a technologist and so this is a tremendous time for all of you to realize that the technology divide has been closed ... the people that understand how to solve a domain problem ... [and have] domain expertise now can utilize technology that is readily available to you. You now have a computer that will do what you tell it to do to help automate your work to amplify your productivity to make you more efficient [...]. The impact, of course, is great, and it is imperative that you activate and take advantage of the technology absolutely immediately and realize that engaging AI is a lot easier now than at any time in the history of computing.*

The combination of your domain expertise in understanding industry problems and the ease of developing solutions with the help of AI opens up a huge opportunity for innovation and staying ahead of the competition. In the next section, we will explain this further.

*The future belongs to the subject matter expert
who can write code, not to tech companies
that don't have your domain expertise.*

Unprecedented time for innovation

Most of the problems in the AEC industry are not new; they have been around for a while. Recent advancements in AI, however, have made tools and technology available to us to solve problems we could not have solved before and open doors to opportunities that were not possible before. That's why this is an unprecedented time for innovation in our industry. For years, you may have waited for tech companies to address some of your problems or develop technology solutions for you. Now you no longer need to wait. With AI, you can leverage your data, which is the most valuable and underutilized asset of your organization, and create new technological solutions for your business.

Take knowledge preservation for example; for as long as we can remember, retaining experts' insights and knowledge has been a valuable goal for many companies. With AI, you can interview the experts in your organization, transcribe their insights, and develop a tool that allows your employees to talk to or chat with and access others' knowledge and expertise. Or imagine documenting and transcribing all the retrospectives and lessons learned when your projects are complete. You can collect data and easily develop no-code solutions to help your project managers better plan and manage projects by identifying patterns in what has been working or not working in your projects.

These are just a few examples. There are countless opportunities to leverage your data and AI to achieve better outcomes for your clients and your organization. But you might ask, "How do we know that the opportunity we're going to pursue will not be developed by tech giants such as Autodesk, Bentley, or other technology providers a few years from now?"

To answer this question, you should understand the differences between the market sizes of the various business improvement opportunities for which you want to develop solutions. In the world of venture capital, we use the term "total addressable market" (or TAM), which is a measure of the total revenue opportunity available if you can build a solution for that opportunity and sell it to all companies with that problem. For example, let's say you want to vaccinate the entire population of the world for a new disease, and you can sell the vaccine for $100. Your total TAM would be $100 x 8 billion people = 800 billion dollars.

But why is knowing the potential TAM important?

If you look at the problems or opportunities that exist in our industry, we can divide them into three categories, as illustrated in Figure 2.1: common industry problems or opportunities that every company has (e.g. improvement opportunities related to finance and marketing), sector-specific opportunities (e.g. opportunities related to stadium seating design or hospitals' patient room design), or company-specific (e.g. you have a secret sauce or methodology that no other companies have). Solutions that address common problems and common opportunities—ones experienced by every company in the industry—have huge TAMs that can run to billions of dollars. On the other hand, solutions that address sector-specific problems or opportunities might have smaller TAMs—more like millions of dollars. Your company-specific opportunities have the smallest TAMs; if the opportunity is truly unique to you, then most likely only you, or maybe a few other companies in the market, need that solution.

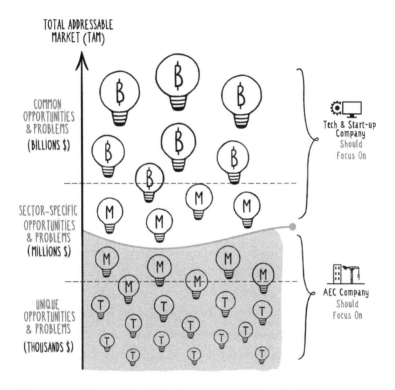

Figure 2.1: Types of business opportunities based on TAM

When we advise technology companies and startups, we always recommend focusing on the top section of Figure 2.1: opportunities with a huge TAM that address common industry-wide problems. This means that the tech company or startup can develop a single solution but sell it to many companies. When we work with AEC companies, we suggest focusing on the bottom section of Figure 2.1: opportunities unique to you. A lot of AEC companies make the mistake of allocating their resources to addressing big industry problems, only to have tech companies or startups come up with a solution that outperforms theirs. The best strategy for you as an AEC company is to partner with startups and tech providers, not build solutions in-house for common industry problems.

*Most AEC companies make the big mistake
of developing in-house solutions for "common" industry
problems. Instead, they should focus on developing
in-house solutions for opportunities "unique" to them.*

To recap, we strongly advocate for software adoption and partnering with startups to tackle common industry challenges. At the same time, recent advancements in AI have made it possible for your company to address unique challenges within your organization. You have the opportunity to invest your time and resources in solving problems and seizing opportunities specific to your needs.

Like the eagle story we told earlier, AI has given every AEC company the wings to soar. Those "wings" are the capabilities to develop solutions for problems and opportunities unique to your company. You can now reach new heights if you put AI to use.

*Are you a technology company that happens to do
design or construction? Or are you a design or construction
company that just uses technology? The choice is yours.*

CONCLUSION

In this chapter, we discussed the disruptive forces of AI from two lenses. First, we looked at how the pace of disruption is accelerating because the time between a technological invention and scaling it across the world is significantly decreasing. Second, the cost of developing software solutions is decreasing. This creates opportunities for subject matter experts in the industry to leverage data, create new in-house tools to address new business opportunities and solve challenges unique to their operations.

In the next chapter, we will discuss how companies are responding to AI and why you should act now before falling behind.

CHAPTER 3

Your Response to AI

There were once four friends living in a vast maze: two mice named Sniff and Scurry and two little people named Hem and Haw. All four of them loved cheese and searched the maze for it every day.

One day, Sniff and Scurry were excited to find the biggest piece of cheese they had ever seen. They began snacking on the cheese every day. But Sniff and Scurry were also wise; they knew that even the biggest piece of cheese wouldn't last forever, so they watched it closely while enjoying it. They soon noticed that their cheese supply was slowly shrinking, piece by piece, day by day.

Eventually, Sniff and Scurry realized they would soon run out of cheese, so they knew it was time to look for more. Before the cheese was entirely gone, they started exploring the maze. Soon enough, their hard work paid off, and they found a fresh, plentiful supply of cheese.

And what about Hem and Haw?

Hem and Haw also discovered a large stash of cheese, and every day became a cheese feast. They became deeply dependent on the cheese but took it for granted and thought it would always be there.

As Hem and Haw continued to enjoy the cheese, they didn't notice it was slowly disappearing. They ignored the signs that their cheese supply was getting smaller and mold was appearing on it, and they didn't think about what might come next.

One morning, Hem and Haw woke up to a shocking discovery: the cheese had disappeared. Somebody had moved their cheese! They were devastated and felt sad, depressed, and betrayed by what had happened. They could not believe their luck had run out. Instead of searching for new cheese, they clung to the hope that it would somehow come back, so they checked out the empty spot day after day to see if it had reappeared. But every day, they left hungrier, disappointed, and weaker.

Nothing lasts forever.

This story was adapted from Spencer Johnson's *Who Moved My Cheese? An Amazing Way to Deal with Change in Your Work and in Your Life.* In his book, Johnson emphasized that change is inevitable, and those who fail to anticipate it are left behind. His story teaches us to strategically seek new opportunities instead of holding on to the past. As the saying goes, "What got us here won't get us there!"

What is the metaphorical cheese in your business? Many AEC executives point to the revenue generated from returning clients as their "cheese." This becomes problematic if your clients are happy with the status quo and resist forward-looking approaches. If your clients are comfortable with the way things are, they might unintentionally hold you back, preventing your company from evolving. These clients pay you based on traditional models like time and materials (T&M) or guaranteed maximum price (GMP), leading you to build your entire business around these models. Much like Hem and Haw, you find yourself stuck in a comfortable yet stagnant room, unaware of how rapidly the world outside is changing.

By leveraging disruptive technologies like AI, your competitors can offer clients better, faster, and more cost-effective services. With the efficiencies they achieve, they can entice your loyal clients away by proposing fixed-price contracts that are significantly lower than what your traditional models demand. Their efficiency not only makes them more profitable but also enables them to manage more projects in a year than you do. In this way, they are not just taking your clients; they are moving the metaphorical cheese and leaving you behind.

---

*Competitors using disruptive technologies
like AI can entice your loyal clients away
with more efficient and cost-effective services.*

Addressing this growing threat depends on how you respond to the sixth wave of innovation. In this chapter, we'll discuss different ways AEC companies are responding to this wave and the characteristics of companies that are succeeding and leading the industry with this transformative technology.

Spectrum of responses to AI

When it comes to AI, AEC companies have responded in various ways. Based on what we have seen in the industry, we have categorized their responses as bystanders, consumers, tinkerers, adopters, and strategic innovators (Figure 3.1). The following sections elaborate on the advantages and challenges of each response.

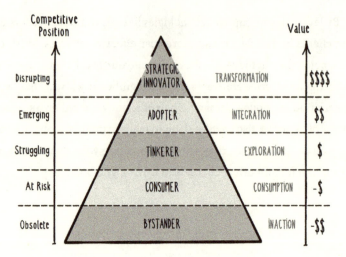

Figure 3.1: Company's response to AI

1. **Bystander:** The first category, bystander, includes the majority of the industry and is made up of companies that take no action and wait to see what happens. These companies often see AI as a hype that will fade soon. There are various reasons these companies take no action. Some work with clients who are happy with the status quo, so they do not see the need to change. Others are led by founders or principals who will retire in a few years and must show a good balance sheet before cashing out and exiting. Some are waiting to see what others will get out of it before taking any action. They see themselves as fast followers who can catch up with the competition later. Some, perhaps like John Antioco of Netflix, don't see a proven return from investing in a disruptive technology like AI and say, "We've been running the business like this for several decades, so why should we change it now?" or "Why change something that ain't broken?!"

   Ravi Bhatia, Business Development Manager at Skanska, said, "Success can sometimes lead to close-mindedness. You think, I've done it this way, so why change? But that mindset can become a real problem."

Regardless of their reasons, these companies have ignored the sixth wave of innovation. They may enjoy the short-term gain and stability from not investing in their company's future, but this enjoyment is unlikely to last long.

2. **Consumer:** Consumers are companies that knowingly or unknowingly use AI solutions in their day-to-day work without fully understanding AI and its capabilities and risks. Every person in the company has a different interpretation of what AI is and what it can do. Due to their lack of knowledge, some employees fear AI and see it as a threat to their jobs, and some are so enthusiastic about its potential that they've developed unrealistic expectations. Moreover, employees may (un)intentionally misuse AI solutions and potentially expose the company to risks, such as proprietary or client data breaches. The inconsistent understanding of AI capabilities and risks can lead to disjointed efforts in leveraging AI solutions across departments and overlooking AI's ethical and privacy risks. Ultimately, these AI consumers could risk their reputation and market position.

3. **Tinkerers:** Tinkerers are the companies that start piloting or experimenting with AI solutions in the market without having a comprehensive strategy. Often, the head of technology in these companies goes to conferences, attends webinars, and talks to peers to find the latest shiny new tools. Or they hear about what other companies are piloting and then try to bring those ideas to their own companies. When you ask these companies' leaders what they are working on, they typically show slides full of the latest experiments they have done and one or two pilot projects that are underway. However, when you ask them how they chose these tools, what the reasoning is behind their project choice, or how these projects align with their company's business strategy, their answers indicate they do not have a comprehensive strategy.

We call their approach an outside-in strategy, which means they are reacting to what's happening in the market without a comprehensive plan. The executives of these companies often ask questions like: "What are our competitors doing in this area?" or "What are the solutions out there we can use?" They just look for tools without realizing that adding more tools without a clear strategy often results in data being locked within those tools, creating data and analytical silos. In practice, the outside-in strategy usually creates tension among team members, as employees are already busy with their daily responsibilities and lack the time to experiment with new tools. This often leads to tech fatigue and exhaustion, leaving technology leaders in a constant position to "push" their ideas and repeatedly struggling to get buy-in from the staff.

4. **Adopters:** Adopters are also piloting or experimenting with AI solutions in the market. However, the main difference between them and tinkerers is that they have an inside-out strategy. They first look internally to find out what business opportunities or challenges AI can address and then look into the market to find out which tool could potentially help them. Instead of "pushing" things to employees, adopters create "pull" from their employees.

One of the challenges adopters in our industry face is that boards and CEOs see AI as a tactical or technological initiative and therefore appoint technology executives (e.g. CIO, head of IT) to lead these initiatives. The board or executives receive occasional updates on the progress of AI initiatives, but the technology leader often struggles to gain organizational traction due to the misalignment of incentives across the organization. For example, business units and divisions are usually assessed by their P&L (profit and loss), so executives prioritize maximizing time utilization. But this makes it extremely challenging to assign people and resources to AI initiatives and gain traction because everybody is busy.

Consequently, the success of these initiatives relies heavily on the interest and capabilities of the small team of volunteers involved. As a result, these projects tend to be small, incremental proof-of-concept efforts that, even though important in building momentum in the organization, rarely achieve meaningful business outcomes. But if adopters could get senior executives' buy-in to put enough resources and energy into their AI initiatives, with the right strategy they could integrate AI into their organization and achieve efficiency and productivity gains.

5. **Strategic Innovator:** Strategic innovators see AI as a business initiative that transforms their processes, core offerings, and business models or even creates new ones. These companies can deliver higher value to their clients through faster delivery and better quality while keeping within or below budget. CEOs and executives of these companies see this opportunity as a transformational initiative. So, rather than adopting AI tools through a technology-led effort, either the CEO or someone from the operation (e.g. COO) leads this transformational effort with the support of the technology lead. Instead of pushing their ideas to the employees, they create a culture of innovation where employees are excited and bring their best ideas forward. Vince DiPofi, the CEO of SSOE, an architectural and engineering firm based in Toledo, Ohio, and listed as one of the top 100 architectural design firms in *Engineering News-Record* (ENR), said this about his company's AI initiative:

The biggest thing you hope for as a CEO is that when you set a vision, instead of pushing people in the company to buy into it, the people in the organization push you to move that initiative forward. Our AI program helped us create a culture and environment where people are extremely excited, and wherever I go in the company, they want to talk to me about it.

In Chapter 4, we will review SSOE's strategic approach in response to the sixth wave of innovation and AI.

Each response to this wave (AI)—bystander, consumer, tinkerer, adopter, and strategic innovator—leads to different outcomes. Bystander companies often become obsolete, while companies who are just consumers of AI usually put themselves at high risk of legal liabilities and reputational damage because their employees use AI without understanding. Tinkerer companies get small business outcomes from their AI experimentations. Because of their outside-in approach, they struggle to keep up with the latest shiny tools and their executives are exhausted by their employees' resistance to adopt. By contrast, adopters create a well-defined strategy, then, depending on the type of opportunities available (as explained in Chapter 2) they bring it to reality either by piloting solutions in the market or developing solutions in-house. By integrating AI into daily operations, adopters emerge as efficient and productive organizations. Unlike adopters who focus on tactical initiatives, strategic innovators perceive AI as a transformative opportunity. Strategic innovator companies foster an innovation culture, challenge their existing business models and operations, and transition into data-driven organizations.

Companies face a choice in the AI wave: stand by and risk obsolescence, consume blindly and increase risk and vulnerability, tinker reactively and struggle, or adopt purposefully for tactical improvements to emerge as an efficient and productive organization. True transformation belongs to the strategic innovators who leverage AI to distinguish themselves in the market and disrupt the status quo.

Who are strategic innovators?

Strategic innovation is the process of identifying and implementing systematic approaches that disrupt or enhance current business models, products, or services to gain a competitive edge and drive long-term growth (17). Strategic innovation involves fundamentally rethinking how an organization creates value, competes, and serves its customers, which often creates entirely new markets or redefines existing ones.

But what are the characteristics of strategic innovators?

One of the characteristics of strategic innovators, and a major advantage, is their ability to capture and utilize the expertise and insights of their employees as data. All employees and teams understand the importance of data. They understand why and how they should contribute to their data and AI ecosystem and that data is part of delivering their projects to the company. In other words, besides meeting project requirements for clients and stakeholders, they also deliver valuable data to their company.

Gradually, with the help of AI, their data and collective intelligence are organized into a knowledge hub that employees can access. When new employees join the organization, they can quickly learn and adapt by leveraging insights derived from the accumulated knowledge of former employees. In addition, when senior employees or field experts leave the organization, their knowledge remains preserved and accessible. This continuity ensures that the organization's expertise is not lost but maintained, which allows the company to rely less on individual "superstars" or "heroes" to salvage projects. These innovators leverage their teams' collective intelligence by capturing it in the form of data and enhancing it through AI systems to consistently achieve outstanding results.

To make this happen, all the business units must have measurable goals and metrics for data delivery in their projects. AI initiatives in these companies are not a side experiment conducted by technology groups or a few enthusiastic employees but are aligned with the overall business strategy. They represent a disruptive shift toward company-wide initiatives backed by the board, CEO, and all executives to deliver major value to clients. By exceeding their clients' expectations and delivering unparalleled value, these companies create and build trust with their clients.

One of the biggest characteristics of strategic innovators is that the executives create a space for capturing and nurturing innovative ideas within the company. This encourages employees to bring their best ideas forward. No one is afraid of proposing or executing new ideas that might fail because failures are celebrated and rewarded as a learning opportunity.

Employees are excited and inspired to work on what truly matters to them. Instead of being bogged down by tedious work that drains their energy, they focus on strategic client problems and let AI systems handle the grunt work under their supervision. These companies disrupt traditional workflows, and that's how they become talent magnets. Instead of trying to convince people to join their company, they have top talent knocking on their doors, eager to work for them.

These characteristics create a dynamic, innovative environment that attracts top talent and delivers exceptional value to clients, leading to exponential growth and market leadership.

To understand this further, let's review an example of a strategic innovator, Moderna, which was one of the first companies to develop a COVID-19 vaccine. Even though Moderna is a biotech company, its approach to strategic innovation can teach valuable lessons to any industry, including AEC.

Since its inception in September 2010, Moderna has relied on digitization to generate data across its operations—from research and development to manufacturing and commercializing products. As the company takes on more projects and researchers conduct more experiments, it collects increasing amounts of data, which feeds into its AI algorithms. In an interview with Forbes (18), Brad Miller, CIO of Moderna, explains:

> *For over a decade, Moderna has built a massive library of data, which is our biggest asset as a digital-first biotech company. This library has allowed us to create our own integrated, proprietary data ecosystem, organize all data sources, and continuously fuel and refine our algorithms with consistency and precision.*

In 2019, the world was desperate to stop the spread of COVID-19. Many large pharmaceutical companies were scrambling to develop a vaccine and get the world back to normal. While many experts predicted that it would take months or years to develop a COVID-19 vaccine, Moderna took just forty-two days to achieve what typically takes years—creating a clinical-grade vaccine ready for trials (19). But how did they do this?

They leveraged AI using all the data and knowledge they had captured over the previous decade to facilitate the identification of target protein mutations and optimize the mRNA sequences used in the vaccine. As well as using data and AI to develop the vaccine, Moderna used them to accelerate the vaccine's clinical trials and distribution. They leveraged AI to identify optimal trial locations and recruit diverse participants by predicting future COVID-19 hotspots (20). This rapid turnaround was possible because Moderna had already spent years building a robust digital backbone by automating data analysis and predictive modeling (19). Despite years of investment, Moderna's financial success from its COVID-19 vaccine was substantial, with total revenues of $18.5 billion. This was a significant rise from $803 million in 2020 (21).

Moderna's story is an example of strategic innovation, which often involves creating a new product, service, or business model that significantly changes a market. Strategic innovation can help you lead a market, create a new market, or expand an existing market and gain brand differentiation and long-term growth.

What is your response to AI?

Take a moment to reflect on where your company stands today and what is needed to progress to the next level. To help with this, we've provided a list of questions in the companion workbook. By answering these, you can fully assess your current position and identify the next steps to advance.

In the next section, we will discuss the risks of not being a strategic innovator and falling behind competitors who embrace it.

What is your company's response to the sixth wave of innovation? Are you a bystander, consumer, tinkerer, adopter, or strategic innovator?

The danger of not being a strategic innovator

If you don't become a strategic innovator, your business will face three major threats. First, your client's expectations will keep changing. Second, your competitors may get ahead of you. Third, the competition landscape will change. Let's look at each threat in more detail.

- **Your clients' expectations keep changing.** Consider the shift from manual drafting to CAD software in the 1980s. Before CAD, architects and engineers drafted plans manually. The introduction of CAD revolutionized the industry, allowing for faster, more accurate designs and complex 3D modeling. As a result, clients began to expect higher-quality designs that were delivered more quickly and within budget. Firms that resisted adopting CAD lost clients to those that embraced it.

Similarly, AI technologies are transforming your client's expectations. Now clients not only expect faster project completion and higher quality, but also demand that this is achieved within or below budget while considering safety and environmental sustainability mandates.

Ravi Bhatia, Business Development Manager at Skanska, said, "It's daunting ... there's constant pressure to ensure safety while delivering projects faster, cheaper, and better."

To remain competitive, you must adapt to your clients' new expectations; otherwise, you risk falling behind. Becoming a strategic innovator is no longer optional but essential to meet your clients' demands and secure your spot in the market. The CEO of an engineering company that lost a million-dollar contract to a competitor said, "It was the first time in my forty-year career that I saw a competitor take our clients away because they use a proprietary AI-powered solution to deliver the project better, faster, and cheaper. We lost a big contract that we used to win all the time." In other words, another company equipped with AI moved their cheese!

- **Your competitors might get ahead of you.** In the AEC industry, while many firms have successfully adapted to past technological shifts like CAD and BIM, adopting AI poses a different challenge. Unlike CAD or BIM, which mainly involve changes in tools and workflows, AI integration demands a more profound transformation across entire operations.

Responding to AI as a technology disruptor isn't merely about buying the latest software. It's about developing a culture of innovation for your employees to cultivate radical business opportunities; preparing your business model to create more value for your clients, and improving your operating model to standardize your processes, improve data quality, and prepare your company for the future of work. That's what strategic innovators do.

Unlike other companies, they focus on taking advantage of this opportunity and positioning themselves so that they can offer clients better, faster, cheaper, and more sustainable services. By doing so, they gain competitive differentiation that reduces market share for competitors that don't act. Emma Viklund is the innovation lead at Skanska, one of the largest construction and development companies in the world. In an interview, she pointed out new possibilities with AI in the AEC industry:

If you don't do anything different, where will you be when everyone else has implemented AI solutions? The pace of change is rapid, and sticking to old methods means being left behind. Just like recycling companies are shifting from waste handling to resource management, construction companies should transit their business; otherwise, not transitioning is moving toward your own demise.

- **The competition landscape will change.** Traditionally, at the national level, the AEC industry has been dominated by well-established firms with decades of experience. However, AI advancements enable new companies to enter the market and reshape the competitive landscape. By leveraging AI, smaller and more agile AEC companies can develop deep expertise and compete with larger AEC firms. In addition, tech startups can also enter this market and offer AI-powered building design services that could potentially disrupt traditional architectural firms by providing faster, more cost-effective design solutions to their clients.

  Moreover, AI can change the merger and acquisition landscape, where AEC companies can acquire other AI startups in AEC to expand their capabilities in the market. For instance, general contractors and builders could acquire AI-driven design startups to provide end-to-end solutions, offering both design and construction services. As Josh McDowell, principal and board member at Mackenzie, an integrated design firm with services in architecture, interior design, and landscape architecture based in Portland, Oregon, highlighted, "AI is such a big game-changer. And if we, as a group, as a profession, don't work together, someone else is going to figure it out, come in and take over."

In short, AI will change client expectations, competitive landscapes, and your market position. Companies that are not strategic innovators risk losing their clients and falling behind, ultimately threatening their long-term growth and success. As Dave Gockel, the CEO of Langan, an engineering and environmental consulting company and one of the top fifty on ENR's 2022 500 Design Firms, pointed out: "Those in the AEC industry who fail to embrace AI and incorporate it into their project delivery systems will find themselves in a never-ending game of 'catch-up' behind their competitors."

If your company does not become a strategic innovator, your long-term survival is likely to be challenged. Other AEC companies or competitors will strategically innovate and can disrupt your company. These innovators can meet evolving client expectations and stay ahead of competitors. Such companies leverage AI to deliver better projects more quickly and on or below budget. This strengthens their client relationships and allows them to win more business. As Bec Wilder, CEO of Green Canopy NODE, a vertically integrated construction technology firm, explained, "Companies that leverage AI will scale faster with higher-quality work—including less rework—and better visibility into cost, which will help contain costs across the board."

Understanding the importance of being a strategic innovator is just the beginning. In the next section, we will discuss what it means to be a strategic innovator and how it can fundamentally change your company's future.

Becoming a strategic innovator

Now that you know what strategic innovation means and why it is important, let's discuss how companies can position themselves in that role. In this section, we outline the main pillars of strategic innovations: culture, business model, and operating model. In the following sections, we will explain each part of the strategic innovation model (Figure 3.2) and their contribution to the overall picture.

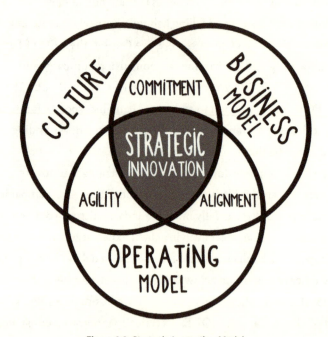

Figure 3.2: Strategic Innovation Model

Culture as the foundation

Becoming a strategic innovator starts with establishing the right culture within a company. You need to encourage innovation, embrace change, and value creative problem-solving approaches that drive disruptive ideas. You and your leadership teams need to provide an environment where employees feel empowered to take risks, challenge the status quo, and collaborate across teams. When you prioritize a culture of innovation and forward-thinking, your company can become a leader in driving changes and position itself as a disruptor in the industry. In Chapter 4, we will explain the steps you need to take as a leader to establish a culture of innovation in your company.

Business model as a value creator

Once you establish a culture of innovation in your company, your teams will bring new business opportunities and innovative AI ideas forward for implementation. However, before acting on these AI ideas, you need to fully understand their impact on your business model so that you can prepare your clients and business accordingly. For instance, if you use an AI tool that makes your design process much faster, you might not want to charge by time and materials. In Chapter 5, we will explain two types of business opportunities and how they could impact your business model.

Operating model as an enabler

To become a strategic innovator, you should align your data, processes, and people. For consistent data capture across all processes, you need to standardize workflows. Your employees should be educated about the value of data, as high-quality data is the key to developing or utilizing AI solutions. Moreover, a successful operating model includes mechanisms for measuring and tracking the performance of your AI initiatives toward becoming a strategic innovator. In Chapter 6, we explain each of these components, as well as the AI team structure, in more detail.

Now that you have learned about the three pillars of the strategic innovation model, let's discuss the intersection of each pillar.

Commitment

Commitment exists at the intersection of culture and business model (Figure 3.2). A strong commitment aligns the company's strategic goals and business model with its cultural values and is defined by the deep dedication of both leaders and employees to the company's long-term vision of becoming a strategic innovator.

Alignment

Alignment is the intersection of strategy and operation, and it tracks whether your business decisions and daily activities support your long-term vision (Figure 3.2). To become a strategic innovator, you should have your people, technology, and processes coordinated with your company's strategic goals. The alignment translates your strategic plans into actions essential for short-term success and long-term growth.

Agility

Agility, in the intersection of culture and operation, enables your company to adjust to changes quickly when needed (Figure 3.2). When operations are agile, your processes are flexible and your employees can pivot in response to new challenges or opportunities. However, you should create a culture promoting agility and adaptability in response to new market trends and demands.

To become a strategic innovator, you need to align and integrate your company's culture, business models, and operating models. This allows employees to feel empowered to share new ideas, run experiments, and scale them across the organization when applicable. You can guarantee the company's growth and long-term success in such a dynamic and agile environment. This is where you can transform ideas into impactful and disruptive AI solutions and become a strategic innovator.

Why act now?

Have you ever explained something to someone, only to hear them say, "Yes, but..." When it comes to innovation, disruption, and change in the AEC industry, it is normal for people to say:

- "Yes, but we've done it like this for many years. Why should we change now?"
- "Yes, but we are a small company."
- "Yes, but our clients are not asking for this."
- "Yes, but most of our projects are with the government and based on time and material."
- "Yes, but we are busy."
- "Yes, but we've heard that before."
- "Yes, but it won't work for us."
- "Yes, but we don't have any data."
- "Yes, but the ROI of this is unknown."
- "Yes, but..."

What are the "Yes, but" statements you've heard before?

Many CEOs and executives see AI as a tactical initiative to improve efficiencies and productivity in their companies. They see it as a tool for incremental improvements.

Our view is that AI isn't just a tool; it's a new way of doing business that affects a company's culture, business model, operating model, client interactions, and even the nature of its services. It is a transformation initiative that impacts every employee in your company. As Jay Edwards, president of Parkhill, an architectural engineering company based in Texas, highlights, "AI has the potential to allow our industry to provide even more innovative, inspiring, and cost-effective solutions to communities by synthesizing more data, analyzing more options, and engaging in broader collaboration."

You should act now because it takes time to build the internal skills required to compete in the age of AI. Taking on this journey sooner rather than later will help you hire the right people with the right skills, acquire better companies, and develop your internal competency. Kodak didn't fail because they were unaware of digital photography—they literally invented it. One of the reasons for their downfall was a lack of skills needed in the digital age. They continued hiring chemists when they should have hired software developers to compete in the digital photography space alongside technology companies. Moreover, being in the film business, Kodak lacked the knowledge to build a profitable business model in the digital space. By the time they recognized the shift, it was too late to develop that expertise.

Kodak's downfall wasn't ignorance of digital photography but the failure to adapt. They chose chemists over software developers in a world rapidly shifting to digital. How is AI changing who you hire?

In addition to hiring the right talents, you should also prepare your people and stakeholders for transformative change. For example, what most people do not recognize in the Netflix and Blockbuster story is that Blockbuster did try to compete with Netflix by launching online services and eliminating late fees. However, this move faced backlash from stakeholders. On the one hand, store owners viewed the change as threatening their businesses, thinking, "The streaming service is going to replace us," so they were completely against it. On the other hand, investors and shareholders saw the $400 million expense as a red flag and a threat to the company's financial stability. This tension led to a dispute over the CEO's salary and, ultimately, resulted in the appointment of a new CEO who canceled John Antioco's plan. The point is that Blockbuster did not lose to Netflix because it lacked technological development; they actually invested in it. Their failure was in not recognizing that streaming and the Internet were transformational initiatives that required them to prepare people and onboard stakeholders within the organization.

*Your ability to leverage AI depends
entirely on how well you prepare and bring
along your people and stakeholders.*

Like the Internet, AI is also a transformational technology. This transformation is not a quick and simple upgrade—it's a complex, strategic shift that requires vision and long-term planning. In our work with AEC companies, we've guided executives in integrating data and AI into their five-year strategic plans. By investing in this forward-looking approach, these companies position themselves to be industry leaders in years to come.

Think of this journey as a marathon without a finish line. You cannot sprint a marathon. You must plan carefully and undergo disciplined training to succeed.

If you are not already running, you are already late.

*Have you considered placing data and
AI at the forefront of your long-term
strategy? Your competitors already have.*

CONCLUSION

In this chapter, we discussed five different approaches companies take toward the sixth wave of innovation: bystander, consumer, tinkerer, adopter, and strategic innovator. Bystander companies struggle to stay competitive, while consumers use AI without understanding and risk their reputations. Tinkerer companies, with an outside-in strategy, exhaust their people by pushing the next shiny new technology to their frontline employees. While adopters see AI as a tactical initiative to gain efficiency and productivity, strategic innovators see it as a transformational opportunity to deliver more value to their clients. Being a bystander and consumer of AI solutions is extremely risky for companies for several reasons. Clients' expectations are changing, and they demand more in a shorter time. Also, the competitive landscape may shift as new players enter the domain. In addition, competitors who see AI as a transformational opportunity may get ahead of you.

In this chapter, we also discussed the importance of strategic innovation and why it is important for a company to become a strategic innovator. To become a strategic innovator, you need a culture of innovation as your foundation, a strong business model that creates value, and an operating model that enables you to execute your strategy successfully. In the next part of the book, we will explain these three main pillars of strategic innovation in more detail.

PART 2

Pillars of Strategic Innovation

In this part of the book, we will discuss how to become a strategic innovator by creating a culture of innovation and preparing your business and operating models for the age of AI.

We will start with culture as a foundation and explain why executives and employees must have a growth mindset in their organization. Next, we'll talk about your employees' and executives' emotions in response to AI, and how to change them from fear to excitement. Then, we will discuss how to create space for collaboration and innovation in your company.

Once you establish a culture of innovation, your employees will bring many business opportunities and ideas. Additionally, we'll explain how these business opportunities impact your business model so that you can prepare your business for the age of AI. You will learn the potential for offering new services and creating new revenue models to set your business apart from competitors.

Once you have prepared your business model, it is time to prepare your operating model for executing these opportunities. We will explore how to manage your data to maximize business value while minimizing risks. We will also highlight the importance of process standardization to ensure consistency and efficiency across your operations. Moreover, you will learn about structuring your AI team in a way that supports strategic innovation with roles and responsibilities. Finally, we will discuss the importance of metrics in measuring progress and leveraging network effects across your company to scale innovation without adding unnecessary complexity.

At the end of each chapter, we'll share a case study based on a visionary company that has successfully implemented the framework, allowing you to see how the concepts discussed are being put into practice. These real-world examples will provide insights into how businesses are transforming themselves into strategic innovators.

CHAPTER 4

Culture as
a Foundation

We had just landed in town to deliver our AI Foundation and Innovation workshop to a client. The day before hosting the workshop, we joined the welcoming and networking event to spend more time with the client and its board. When we arrived at the venue, it was set with multiple round tables in banquet style. We were shown to a table where the leadership team sat with their board members. After a short introduction, Rose, who had been a board member for over eight years and had more than four decades of experience as the CEO and president of a construction company, looked at us and asked, "So, are you going to host an AI workshop for us tomorrow?"

We responded with excitement, "Yes, are you ready?"

The expression on her face didn't change, and without blinking, she continued, "You know what, this thing ... AI ... is scaring me. I'm worried about the next generation and whether they will have jobs! It looks like AI will replace them."

It was not the first time we had heard this concern, and we knew it would not be the last. In response, we briefly explained AI and how it could help us to do better and more with less. We promised her we would dive deep into fundamentals during tomorrow's AI workshop.

Rose smiled at us and responded: "Let's see ... I look forward to your session!" From the look on her face, we could tell she felt skeptical about AI and didn't believe what we said.

With the networking event over, we returned to our hotel. We couldn't wait for the next day when we would see everyone at the AI workshop and continue our conversation.

On the day of the workshop, the room was packed with about a hundred people. There were people from the executive teams, board members, and a few employees. Throughout the workshop, we discussed AI, what it can and cannot do, and a few industry use cases. We also reviewed the AI strategy framework from the book *Augment It* and how AI can be implemented strategically. Once the participants had learned more about AI and previewed use cases relevant to the AEC industry, they completed some hands-on activities to familiarize themselves with how AI can help them and their teams in their day-to-day jobs. This exercise gave everyone a practical understanding of AI's transformative potential in driving the company's future business and growth.

By the end of the workshop, everyone was excited about AI and asking questions about how to bring its value to their teams and operations. We experienced a lovely, heartwarming moment when Rose came to us and said, with a smile on her face and a light in her eyes, "You know what ... you CONVERTED me!" Her expression made our day, and we felt we had managed to change her perspective and potentially that of others who might have felt the same about AI.

But what do you think has really changed here? Why was a board member, who was concerned about AI the day before, now excited about its potential?

The only thing that had changed was that she had been educated— she had learned about AI and its transformative outcomes. So, Rose could see AI as an augmentation, not a replacement.

Sometimes, we wonder how many board members, CEOs, and executives in AEC companies have the same thoughts and feelings as

Rose. And with those negative emotions and mindsets, how can they lead their companies to survive during the sixth innovation wave?

These negative emotions can create barriers to forward thinking and innovation. However, from our experience working with many AEC companies, we have seen that those companies that build the innovation culture from the outset are better equipped to overcome these challenges and thrive in the face of disruption. In other words, to become a strategic innovator, you should start with culture and build a foundation for innovations. In this chapter, we will discuss the practical principles of creating a culture of innovation in your company (Figure 4.1).

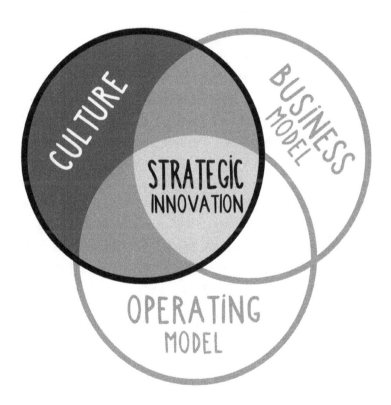

Figure 4.1: Culture as the foundation

Create a culture of innovation

Innovation culture is a set of shared values and practices that encourage innovation, embrace change, and prioritize creative problem-solving to drive novel ideas. Your role as a leader is to provide an environment where employees feel empowered to take risks, challenge the status quo, and collaborate across your organization. John Kotter, a distinguished Harvard Business School professor, author, and leading authority on leadership and change management, said (22):

> In terms of getting people to experiment more and take more risks, there are at least three things that immediately come to my mind. Number one, of course, is role-modeling it yourself. Number two is when people take intelligent, smart risks, and yet it doesn't work out, not shooting them. And number three is being honest with yourself. If the culture you have is radically different from an experiment and take-risk culture, then you have a big change you are going to have to make—and no little gimmicks are going to do it for you.

Fostering a culture of innovation brings numerous benefits to your company. Creating a culture with a positive attitude to innovation will lead to a more innovative and creative workforce, enabling your company to be more agile in response to market changes and offer unique services. By encouraging your employees to experiment and embrace new ideas, you will stay relevant, meet evolving customer needs, and drive sustainable growth.

But how can you create a culture of innovation in your company?

To create a culture like this in your company, we have outlined the main principles based on best practices and our real-world experience (see Figure 4.2):

- Lean in
- Set budget and ROI expectations
- Shift employees' emotions
- Create a space for innovation
- Encourage internal and external collaboration
- Celebrate failures as learning opportunities

In the remainder of this chapter, we'll discuss each principle in detail.

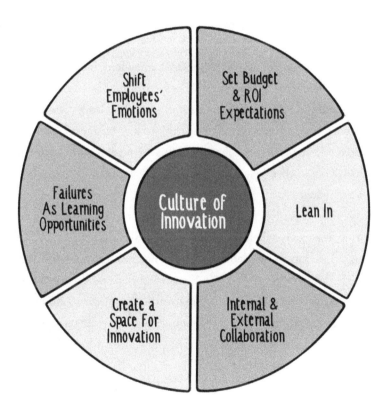

Figure 4.2: The main principles of the culture of innovation

Lean In

In almost every discussion we've had with CTOs, CIOs, and IT directors of AEC organizations over the past several years, a recurring theme has been: "Our CEO or board is not fully supportive; how can we get their buy-in?" They are absolutely right—this is often a big blocker to action. Without buy-in from CEOs and boards, no company can create a culture of innovation. Conversely, we have also encountered companies where the CEO was excited about AI, but their board and leadership team see AI as a distraction to their business. In this case, CEOs are responsible for setting the tone, emphasizing the opportunities AI can bring to the company, the risks of not acting, and getting everyone's buy-in.

If your company doesn't have a CEO, president, or someone who oversees the entire organization, it's essential to assign at least one senior leader to take charge of this initiative. This leader should ideally have a strong background in operations and business to guide AI initiatives strategically. For the rest of this chapter, whenever we refer to the CEO and their role, think of the equivalent senior executive in your organization.

While everyone is responsible for this journey, CEOs play a critical role, especially when deciding to undertake strategic innovation. Often, AI initiatives are dedicated to the head of technology without the direct involvement of a CEO. However, because strategic innovation initiatives with AI often require creating new business models or changing current ones, entering new markets, or serving new clients, these initiatives must be CEO-driven. CEOs have a holistic view of the company's long-term goals and can align these initiatives with broader strategic objectives. Without a CEO's commitment and direct involvement, strategic innovation initiatives using AI won't succeed. As Vince DiPofi, the CEO of SSOE, explains:

As CEOs, we must be selective in what we focus on, but leading AI initiatives should be a top priority. AI touches on every aspect of the business—data, privacy, ethics, clients—making it a truly transformational force. By driving these initiatives ourselves, we signal to the entire organization just how critical they are. In my experience, when the CEO takes the lead on something, it sends a powerful message about its importance. AI is one of those rare opportunities that can reshape not just our companies but the entire industry.

Having said that, we do not suggest that CEOs should be involved in day-to-day operations, attend all the meetings, and carry the burden of execution in this process. CEOs and executives must, however, commit to becoming advocates, prioritizing these initiatives for the entire company. They should create a vision and mission for the initiative, clearly articulate it with employees and the board, and ensure that adequate resources exist. This process takes time and requires a lot of effort. Changing organizational culture does not happen overnight. Brad Miller, CIO of Moderna, explains (18), "Stephane [Bancel], our CEO, is the number one advocate for AI, as we really see that AI has been, and will continue to be, the way in which we will transform our company."

While the head of technology can often manage AI initiatives without direct CEO involvement, strategic innovation with AI requires CEO-driven leadership. CEOs must align these transformative initiatives with the company's long-term goals to ensure success, making their commitment crucial for strategic innovation.

Set budget and ROI expectations

Most CEOs of AEC companies ask how much the budget for their innovation should be in the age of AI. This is a great question and a very difficult one to answer. In a *Harvard Business Review* article, Gina O'Connor discusses several industries and the percentage of revenue that companies typically invest in R&D. Some examples are technology (15—20% of revenue), pharmaceutical (15—25%), automotive (5—10%), and consumer goods (3—5%) (23).

What about the AEC industry? Do we have a proper R&D budget similar to other industries? According to McKinsey's report (24), the average IT budget in the AEC industry is around one to two percent of a company's total revenue, which is much lower than many other industries. Some companies include their R&D budget in this number, and some don't. Some companies don't even have an innovation budget. In this case, we recommend starting with one percent of your total revenue. After the first year, when you have actual data, you can adjust your innovation budget accordingly. Abul Islam, the CEO and president of AI Engineers, a Connecticut-based multidisciplinary company providing civil, structural, wastewater, and building systems services, explains:

Many big companies isolate their technology groups, focusing only on monthly profits, and people just keep doing the same thing. That's why I'm betting the industry will change by investing in R&D. We invest between one and two percent of our total revenue in research and development. I encourage other CEOs to do the same.

Start with 1% of your revenue as your innovation budget and adjust it based on your first-year data.

While allocating enough budget for strategic innovation with AI is already challenging in our industry, some CEOs and board members set unrealistic ROI expectations for AI initiatives.

The head of innovation of an engineering company reached out to us and shared his struggle to find the right AI opportunity for his company. His challenge was that he could not guarantee the ROI of the AI opportunities to his CEO and board. Over the course of a year, he explored more than 200 AI opportunities but could not guarantee any definitive outcomes. When we recommended that he stop calculating ROIs and start projects on a smaller scale, he responded, "You don't understand; our CEO is very ROI-focused. He just doesn't allocate budget without an immediate return." He was in analysis paralysis mode! We're not against ROI calculations in business; in fact, ROI is an excellent metric to assess the financial viability of investments, especially short-term investments in problems and solutions where outcomes are relatively predictable and quantifiable and happen within a clear timeframe. That said, it is very difficult to predict the outcome of strategic innovations because they are long-term, transformational initiatives with unknown and unpredictable outcomes. Do you think Jeff Bezos knew Amazon would develop AWS when he started the company? Did Steve Jobs have a plan for an app store when he was working on the iPhone? Do you think they calculated the ROI of their investments before starting Apple or Amazon? Of course not—it was not quantifiable.

In your case, what if implementing one of your ideas creates a new business model or opens up a whole new market segment? How do you quantify that?

These strategic innovation initiatives have intangible outcomes that are difficult to quantify and take time to produce ROI. For example, where do you think top industry talent will want to work in the future? At an innovative firm with a culture of innovation at its core? Or at a traditional AEC company where ideas are overlooked, or no one has time for creativity and innovation?

In the long run, strategic innovation positions your company as a leader in the industry and attracts lots of new talent to your company. How do you measure the ROI of having the best industry talent working for you?

In his book *The Innovator's Dilemma: When New Technologies Cause Great Firms to Fail* Clayton Christensen highlighted (25):

> *Companies whose investment processes demand quantification of market sizes and financial returns before they can enter a market get paralyzed or make serious mistakes when faced with disruptive technologies. They demand market data when none exists and make judgments based on financial projections when neither revenues nor costs can, in fact, be known.*

Calculating ROI without considering other factors can lead to a company's long-term failure. Perhaps Kodak's executives calculated the ROI of their investment in digital photography, but the numbers did not add up, and this ultimately cost them their business. Not investing in strategic innovation opportunities will, in the long run, be far more costly than the initial investment. Your competitors have already started this journey, and by the time you decide to play catch-up, it might be too late or too costly. Remember the Netflix story discussed earlier? After declining to acquire Netflix for $50 million, John Antioco's Blockbuster spent $400 million trying to catch up with them. But it was too late, and Blockbuster went out of the game.

Instead of focusing only on measuring the ROI, forward-looking companies in the AEC industry prioritize the long-term value of innovation. For example, Andy McCune, CEO of Wade Trim, explained, "If you're a CEO who wants to see a financial calculation, you're being short-sighted; you're not taking the long view." With this mindset, he established a culture encouraging his employees to look for the risk of not investing (RONI) in innovative ideas. Tim O'Rourke, the CTO of Wade Trim, explains:

> *Within the Office of Applied Technology (OAT) at Wade Trim, we have a process for capturing and evaluating employees' innovative ideas. Besides the conventional metrics, such as ROI, we also evaluate ideas based on the risk of not investing in them. If we don't invest in novel ideas, especially around what differentiates us in the market, we may lose our revenue and market share in the future.*

Calculating short-term financial returns is incompatible with strategic innovation, which is typically a long-run initiative. Because traditional ROI is based on quantifiable financial returns, focusing on it may discourage investments in strategic innovation where the benefits are less certain or more complex to measure.

The best approach in measuring the return on investment of your AI initiatives is to use metrics other than immediate financial return, such as the percentage of ideas aligned with the company's long-term strategic goals, competitive positioning, talent attraction, customer value, new knowledge and organizational learning, cultural impact, and thought leadership. By adopting these broader metrics, your innovation initiatives may not create short-term gains but will be strategically aligned with your long-term vision.

Traditional ROI discourages strategic innovation, which requires long-term metrics like strategic alignment, competitive positioning, and cultural impact, rather than short-term financial returns.

Shift employees' emotions

When you're reading news articles, browsing your favorite social media, or watching the news on TV, do you feel encouraged and empowered, or do you feel a bit depressed? We mostly feel the latter. This feeling of depression is consistent with the findings of Dr. Deborah Serani, the award-winning author of *Living with Depression*. Dr. Serani is also a licensed psychologist with over thirty years of practice, and a senior adjunct professor at Adelphi University, located in Garden City, New York. She explains (26), "Fear-based news stories prey on the anxieties we all have and then hold us hostage. Being glued to the television, reading the paper, or surfing the Internet increases ratings and market shares—but it also raises the probability of depression relapse."

Knowing this is important because most of your employees have heard or read about AI in the news or on social media, which are often unreliable sources of information. Most likely, many employees in your organization don't know what AI really is, and they're afraid of being replaced by it. Hence, they are reluctant to adopt it or participate in any data or AI initiatives.

We have learned this through the innovation workshops we have hosted over several years. During the workshops, we often survey people's emotions at the beginning and again at the end of our AI sessions. Once, we cross-analyzed more than 1,200 responses to look for patterns. The top five emotions reported at the beginning of the workshops were fear, confusion, uncertainty, skepticism, and curiosity.

Similarly, your employees, whether senior leaders, board members, or grassroots-level staff, may experience these emotions, but they are unlikely to openly discuss them. If these uncomfortable emotions are not addressed, your AI innovation initiatives will not succeed.

We were once approached by a consulting company executive who shared that their biggest AI investment in the company had failed. They developed a model-based estimating tool that was not adopted by the company's estimators. This was because the estimators were fearful about AI and its potential to replace them. Unsurprisingly, they did not want to use it. Then, when management forced them to use it, they deliberately put the wrong data into the system and this eventually caused the AI project to fail. This is why you should start by changing people's emotions from negative ones (e.g. fear and anxiety) to positive ones (e.g. excitement and empowerment).

How can you change people's emotions?

Earlier, we mentioned the surveys we run during our innovation workshops and how employees' emotions shift from negativity to positivity. But what do you think changes between the beginning and end of the workshop?

The participants have been educated!

Anuj Aggarwal, Director of Deep Learning and AI at NVIDIA, highlights the value of education: "As advancements increase across the AI landscape, education is an imperative that must be considered throughout the use and implementation of AI. Without proper education throughout the process, large amounts of time and resources could be spent with unexpected outcomes."

Education and training transform your employees' fear into empowerment.

Education and training help your employees and executives better understand the field of AI and, therefore, overcome their fears. With proper education, they will realize that adopting AI does not mean their company needs fewer employees but means that the company can do more with less. This approach will make them more likely to support your initiatives than resist them or cause them to fail. In other words, the starting point of your AI journey is preparing people within your organization. As Keith Tencleve, Director of Innovation at Garver, a multidisciplinary engineering, planning, and environmental services firm, shared, "As you embark on your AI journey, you need to figure out how to develop your pipeline of people who can think in the way today's technology requires because we're a people-powered business."

Create a space for innovation

In the 1940s, William L. McKnight, the former president and chairman of 3M, believed that the company needed more than formal research and development processes for its growth and success. To that end, he initiated the 15% Rule. His goal was to allow every employee to explore and experiment with new ideas without being constrained by their day-to-day responsibilities. Under the 15% Rule, 3M employees could spend up to 15% of their working time pursuing innovative ideas they were passionate about. This policy created a culture of innovation and resulted in many successful products, such as Post-It Notes (27). The 15% Rule later inspired similar initiatives at other companies, including Google's 20% Time, which led to innovations like Gmail and Google News.

When speaking with CEOs and leaders of AEC companies, one of the biggest challenges they talk about is their employees having no time for ideation and innovation because they are busy with a backlog of projects. Moreover, since most of these companies operate on a billable hour model, which in some cases accounts for 95% of their employees' time, allocating time for innovations means less revenue. Their dilemma is how to choose between maximizing billable hours to increase revenue and satisfy shareholders, even for the short term, and allocating a percentage of employees' time to innovation and future growth.

An effective way to allocate time and encourage innovation is to create an internal project where employees can submit their ideas and log the hours they dedicate. This enables you to track both participation and time spent. To enhance this, consider introducing an innovation grant to support and reward new ideas. Michele Stanton, the CIO of

HGA, a national interdisciplinary design firm rooted in architecture and engineering, explains:

To support innovation opportunities within our project delivery practice, we established an innovation grant process and budget. Project teams are encouraged to apply for innovation grants throughout the year, which are reviewed and approved in just a few days. With this approach, project fees don't need to take the hit of innovation research and development, enabling teams to generate innovative solutions driving even greater impact in their projects.

Creating a space for your employees to ideate, collaborate, and innovate might initially reduce billable hours as they need time for researching and brainstorming. However, this investment will boost your employee engagement and retention rate, ultimately driving your long-term business growth. The value gained from these innovations and the resulting growth will far outweigh the short-term trade-off in billable hours.

Provide space for employees to innovate!

Encourage internal and external collaboration

While leaning in and providing a safe space for your employees to ensure the success of your initiatives, you should also encourage cross-organizational collaboration among employees. Often, the best ideas come from people unfamiliar with a topic's constraints. By encouraging cross-organizational collaboration in your AI innovation initiatives, you allow diverse perspectives and expertise to come together, leading to more creative solutions and faster problem-solving.

In one of our ideation sessions, a group of senior leaders discussed the company's new business model and brainstormed how to position their services in front of their clients. Most of the senior executives who had been in the company for over two decades agreed to provide more discounts to their clients to win more jobs. But a new marketing hire, with a lot less work experience and knowledge of the business, suggested a different approach: increasing the price by adding new value to their packages and expanding the target market to include new clients. Despite being new to the company, her idea was recognized as a brilliant strategy worth exploring to see how it could help the company stand out and attract new clients. "This was so out of the box," the senior executives said.

In addition to cross-organizational collaboration, you should consider bringing new partners (e.g. consultants, university collaborators, and other AEC companies) into your innovation ecosystem. With their cross-industry knowledge, they could update you on industry patterns and add tremendous value to your initiative.

Procter & Gamble (P&G) is an excellent example of a company that taps into external expertise and ideas instead of solely relying on internal research and development to stay competitive in the market. P&G's Connect + Develop program invites ideas and innovations from outside the company, enabling the co-development of new products and solutions. For example, Swiffer cleaning products and Olay skincare emerged from this program. This strategy created over 35% of the company's new products and billions in revenue that originated from external sources (28).

Like P&G's Connect + Develop program, which leverages external expertise to offer new products, you can also benefit from collaborating with external partners.

Take a moment to reflect: Who are your external collaborators for innovation—those who can help broaden your perspective? Are there collaborators you should have but don't yet?

Celebrate failures as learning opportunities

Imagine if Thomas Edison had given up after his first few hundred failed attempts at creating the light bulb. What if he had accepted those failures as proof that he wasn't capable of success? Fortunately, he didn't. Instead, Edison's belief that his abilities could improve with effort led him to persevere through 2,774 failed attempts at inventing the light bulb (29) until he eventually created one of the most transformative innovations in history.

You might wonder what type of mindset drives innovators like Edison to keep pushing forward after so many failed attempts. To explain this, let's consider Jack and Jay, two people who have different mindsets.

Jack believes he was born without specific skills, so he will never be able to learn anything that requires those skills. Since he cannot accept failure, he shies away from unfamiliar things. He is afraid of making mistakes, and he only takes opportunities if he's sure he'll succeed. He is threatened by other people's success and tries to prove himself to others by succeeding in his projects.

In contrast, Jay believes talent is not static and can be achieved by effort. Therefore, she always focuses on her effort and process, not the outcome of her projects. This attitude enables her to take on projects out of her comfort zone. What happens if her project fails? She learns from it and takes that learning to the next project.

While Jack often waits for the perfect opportunity, Jay takes on new projects, learns from them, and grows.

As we said earlier, Jack and Jay have different mindsets. Jack's mindset is called a fixed mindset, while Jay's is a growth mindset. In her book *Mindset: The New Psychology of Success* Carol S. Dweck talks about these two types of mindsets (30):

> *Believing that your qualities are carved in stone—the fixed mind-set—creates an urgency to prove yourself over and over. If you have only a certain amount of intelligence, a certain personality, and a certain moral character—well, then you'd better prove that you have a healthy dose of them... I've seen so many people with this one consuming goal of proving themselves—in the classroom, in their careers, and in their relationships... Growth mindset is based on the belief that your basic qualities are things you can cultivate through your efforts, your strategies, and help from others. Although people may differ in every which way—in their initial talents and aptitudes, interests, or temperaments—everyone can change and grow through application and experience.*

Which mindsets should you establish at your company?

A growth mindset is important because when you start your AI journey, a good percentage of your AI projects may not yield the results you expected from the outset. With a fixed mindset, you may think these are failures and stop the initiative. However, with a growth mindset, you will see failure as a learning opportunity to understand why things didn't work and how to improve things on the next project. This concept might be foreign to many AEC executives because they want to deliver successful projects to clients. So, running innovation initiatives in which failures are celebrated, and often seen as more important than success, is unusual.

We both used to work as structural engineers and know experimentation and failures are not options for a lot of professionals in our industry. When working with engineering firms, engineers can barely accept the idea of experimenting and failing in their AI innovation initiatives—it's just not part of their mindset. And guess what? When these engineers become executives and board members, they hold on to their old beliefs, which makes it challenging to encourage growth mindsets and celebrate failures.

Speaking of failures, we are not suggesting that company executives fail on AI projects due to a lack of diligence or poor execution. On the contrary, we emphasize the importance of careful planning and execution. But we also recommend celebrating the process and effort invested in these projects more than the outcome. And let's be honest; if all your AI initiatives succeed, then your teams probably played too safe and did not push the boundaries of innovation far enough to move your firm toward being a strategic innovator.

Throughout this process, your teams should monitor progress in your AI initiatives. By doing so, you can gain insights into what worked and what didn't and define the biggest takeaways so you can leverage them to further refine your processes and strategies. This is to create a growth mindset for your entire organization and let people take on new challenges to learn and extend their boundaries.

Creating a culture of innovation is not a one-and-done project. It's a journey of continuous improvement for your company.

True disruption comes with risk; failure is a sign that you're challenging yourself and exploring new ground.

CONCLUSION

In this chapter, we discussed how to create a culture of innovation in your company. You learned that CEOs should lean in and support innovation initiatives. You also learned that calculating ROI for long-term disruptive innovation initiatives does not work. Instead, you need other ways to measure the success of your projects. When employees have psychological safety, they are more likely to take the initiative and share innovative ideas, thereby contributing to your success. As a leader, you should have a growth mindset for exploration and development. You should establish that mindset across the organization by creating a culture that encourages taking risks, learning, and continuous improvement. So, celebrate failures as learning opportunities.

Now you know how to establish a culture of innovation as the foundation of strategic innovation. In the next chapter, we will discuss your business model as a value generator for strategic innovation in your organization.

SSOE'S JOURNEY IN CREATING A CULTURE OF INNOVATION

Since 2009, SSOE has delivered more than $2 billion in savings to clients across industrial manufacturing and general building sectors. For the past decade, SSOE has consistently positioned itself as a top-ranked architecture and engineering firm. Since 1948, while delivering unparalleled value to clients, SSOE has also cultivated a workplace culture that earned the Great Place to Work® certification and was named among the "Best AEC Firms to Work For" by Building, Design + Construction. Moreover, SSOE has repeatedly been ranked high across multiple divisions in the Engineering News-Record, showing its strong leadership and performance in different areas of the industry.

We met Vince DiPofi, the CEO of SSOE, and his executive team in San Francisco. From the first interaction, it was apparent that our companies shared the same vision: both saw AI as an opportunity for transformation, not just a tactical business opportunity. The following paragraphs summarize SSOE's journey into creating a culture of innovation from Vince's point of view.

I see AI as a transformational opportunity for SSOE's future. During the 2022 shareholder meeting, I strategically introduced the concept of integrating AI into SSOE's operations, sparking a company-wide conversation. Understanding that AI adoption wasn't just a technical challenge but a cultural one, I knew that to succeed, we should obtain a commitment from all stakeholders—from the board members to frontline employees.

To build this commitment, my executive team and I dedicated six months to understanding AI's potential. We conducted deep research, consulted with industry experts, and discussed the technology's impacts on SSOE's business model, operations, and workforce.

The decision to develop an innovation strategy for the company wasn't just an idea—it was a well-researched initiative that the leadership team and shareholders confidently supported. Moreover, to achieve enterprise commitment and support from all levels in the company, we even rewrote our strategic plan to incorporate AI.

At the beginning of our AI initiative, we faced two challenges: First, we knew that not all employees have positive emotions about AI. In fact, the poll results showed that "anxiety, pessimism, and concern" were among the top emotions before the start of our AI initiatives. Another challenge for us was the lack of buy-in from all executives. Some executives were extremely bullish about this initiative, but others were not. With the right AI training, we could address both challenges at the same time.

Through training, we transformed everyone's sentiments about AI from anxiety, pessimism, and concern to a sense of opportunity, optimism, and excitement. The other primary outcome of the training for our shareholders, leadership team, and employees was understanding the interrelationship between the business process and AI. This was a significant shift from the past mindset, where technology was often seen as a standalone fix rather than a tool that could transform daily operations.

As employees learned more about AI, they began to see how it could enhance their work rather than replace it. Conversations across the company shifted, with everyone discussing business opportunities AI could bring to SSOE. Our approach eliminated fears, enabled executives and teams to see the tangible benefits AI could bring to their daily business, and created consensus and buy-in from all executives from different business units.

Once employees' emotions shifted, it was time to set a space for innovation. We knew we should set a budget for our AI initiative, but we did not know exactly how much. So, we initially allocated a half-percent of our revenue to this. Besides budgeting, we also decided to allow our employees to spend two to three hours per week learning. This enabled our employees to explore AI's incremental and radical potential for the company. By creating this space for ideating and forward thinking, we could position this initiative as a collaborative effort involving the entire organization.

We had several collaborators in our AI initiative, both within the company and with external partners. From operations to business and IT, teams have worked together to identify business opportunities enabled by AI's power. We also tapped into external expertise that complements our internal efforts. This blend of internal engagement and external expertise was instrumental in driving SSOE's cultural shift and has positioned the company for long-term success.

In our AI initiative journey, I knew that innovation lies beyond our comfort zone, where mistakes are inevitable and necessary. I encouraged my teams to push the boundaries of what was possible and see failures as learning opportunities. Before choosing the AI opportunities to shape our AI strategy, I sent a very clear message to my team: "If we succeed in all AI opportunities, then it means we haven't pushed the boundaries far enough." Setting this mindset encouraged teams to explore not only common and essential AI opportunities but also strategic innovation initiatives and take risks. I know we cannot stand out in the competitive market without taking risks and bold steps and being strategic innovators.

I always share the famous quote from Peter Drucker with my team: "Culture eats strategy for breakfast." This is especially true in this new age of AI. As a CEO, you should start with building this foundation first. And the biggest thing you hope for is that when you set a vision, instead of pushing people to buy into it, they push you to move that initiative forward. I'm extremely happy to say we built a culture at SSOE where people are genuinely excited, and wherever I go in the company, they want to talk to me about it.

You won't lose your job to AI; you'll lose your job to a competitor that uses AI.

CHAPTER 5

Business Model as a Value Generator

"This is the race to the bottom!" We've lost count of how often we have heard this phrase from company executives working with their clients on a time and material basis. In this model, the more hours you spend on completing a project, the more money you can usually make, so eliminating mundane and redundant tasks goes against financial logic. An efficiency gain tool like AI can reduce the hours needed to deliver projects, so in this case, the fees earned would be reduced.

We wanted to test this case. Is it really true that using an efficiency gain tool like AI, which can help designers design ten times faster, will reduce the amount of time spent on the project by ten times?

In one of our past projects, we ran a scientific test.

At that time, we were developing an AI tool that could reduce design time by ten times. We wanted to measure how fast designers could design a structural bracket, like the ones you put under a shelf, with specific structural and aesthetic requirements. So, we recruited Jill, a rocket scientist with five years of experience, to use the AI tool and give us feedback. We also had a team of user researchers, designers, and engineers watching and timing every click and move she made. They wanted to test this hypothesis: "Designing with an AI tool that enables designers to work ten times faster than with traditional tools will result in a proportionate reduction in project delivery time."

After hours of testing, we had the results. Contrary to what we expected, Jill took more time to complete the project, not less! Why?!

Even though the tool could design a bracket ten times faster than a traditional tool, Jill did not want to deliver the first design she came up with. She wanted to do more! She used the tool to design and explore more alternatives. Also, she used it to analyze several what-if scenarios and evaluate new constraints that she could not have done without the AI tool.

So, our hypothesis was proved wrong. Gaining ten times greater efficiency with AI does not necessarily mean spending fewer hours on projects; in fact, Jill spent more hours delivering the project because she did a more comprehensive and thorough design and study than she would have when using the traditional tool. In other words, she could deliver more "value" in slightly more time. Imagine if she was working on a client project; shouldn't she charge more based on the new value she delivers to the client?

This is not a race to the bottom; this is a race to the top for the companies that create more value for their clients.

Some might read this and think, 'If I had that tool, I'd just ask the designer to deliver the first design quickly.' While that's possible, it misses the point. The point is, most executives don't realize AI represents a new class of tools with significantly more capabilities and value to clients than conventional tools. Let's consider another example.

Imagine you're in the ground transportation business: your conventional fleet consists of buses and vans, and you have recently acquired a new class of fleet in the form of an airplane. You currently charge your ground transportation clients based on the hours spent reaching their destination. But how should you charge your airline clients? Driving from San Francisco to Los Angeles takes almost seven hours, while a flight takes one and a half hours. So, taking a flight is a lot faster, but should it be cheaper?

Ground transportation and airlines have completely different business models. In other words, you cannot run an airline using the same business model you would use for a bus company. As you look for business opportunities to leverage disruptive technologies (like AI), you should also reimagine your business model because it enables you to significantly change the value you provide to clients.

In the previous chapter, we explained the first pillar of the strategic innovation model: creating a culture of innovation that encourages your employees to bring new business opportunities and innovative ideas forward for implementation and execution. In this chapter, we will focus on the second pillar: the business model, an essential component for generating value (Figure 5.1). But what exactly is a business model?

Figure 5.1: Business model as a value generator

According to the *Harvard Business Review* (31), a business model describes how a company *creates*, *delivers*, and *captures* value. A successful business model creates value for clients, companies, employees, stakeholders, and partners. Value creation involves developing products, services, or experiences that benefit clients or solve their problems. Value delivery refers to how the value created reaches customers through distribution networks, logistics, customer service, and so on. Value capture refers to strategies for monetizing the value delivered to customers, such as pricing.

AI fundamentally changes how business value is created, delivered, and captured. Hence, it is crucial to adapt your business model to leverage the opportunities that AI brings to your organization. Therefore, in the next section, we'll explain two types of business opportunities (incremental and radical) and their impact on your business model. Once you understand how AI will affect your business model, we'll show you how to use AI to race to the top rather than racing to the bottom.

Two types of business opportunities

Generally speaking, there are two types of opportunities for established businesses: incremental and radical (see Figure 5.2). It's important to understand these types of opportunities because they bring different values to your organization. Let's explain what they are.

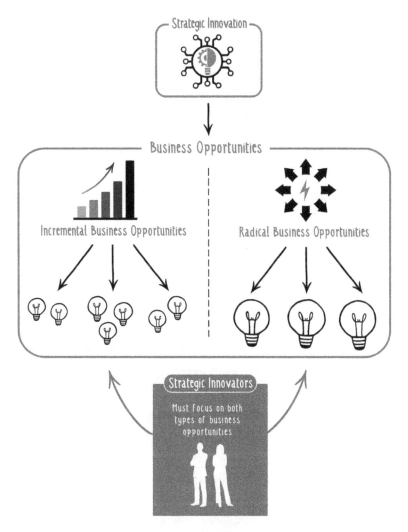

Figure 5.2: Two types of business opportunities

Incremental business opportunities

Incremental business opportunities focus on optimizing and improving your organization's current business operations. These opportunities might be short-term initiatives or quick wins with immediate benefits that enhance efficiency, reduce cost, and solve well-defined problems. An example of an incremental business opportunity is using an AI chatbot to find information across several platforms within your company.

How do you find these opportunities? If you ask people what bothers them the most during their day-to-day tasks, you'll get a long list of opportunities for improvement in your organization. These opportunities can be found in both the business (corporate) side of your company and your operations side. Examples include using AI for: selecting better projects, writing better proposals, projecting your cashflow, better project planning, allocating resources, creating more design options for clients, doing faster engineering analysis and optimization, QA/QC, improving safety, better scheduling, weather predictions, predictive maintenance, and so on. Companies often have a backlog of potential areas for improvement.

Historically, companies focused only on addressing pain points in their organization. But with AI, you can augment your capabilities in the areas where your company excels—your secret sauce. Let's say your company is really good at project estimating because you have a unique process, data, or subject matter expert. You can develop an AI solution to scale this capability across the organization and reduce reliance on experts who might be retiring soon. By doing this, these experts' knowledge and capabilities will stay in the organization when they decide to retire.

While incremental business opportunities are essential, if not selected correctly, they often lead to improvements that could be easily replicated by competitors. For example, when OpenAI released ChatGPT, many AEC companies were interested in how this technology could help them analyze requests for proposals (RFPs), which is time-consuming. This opportunity could be easy to build. But also easy to copy. Once your competitors see your success, they can quickly adopt similar strategies or solutions.

Now let's talk about the other types of business opportunities.

Radical business opportunities

In contrast to incremental business opportunities, which focus on existing markets, processes, and services, radical business opportunities have the potential to reshape and transform your entire business. They are typically about creating new markets or offering new services to existing ones. Either way, they generate new value for your clients and new revenue streams for you, and make it necessary to invent a new business model or reinvent the existing one. While radical business opportunities often involve a higher level of risk due to their novel nature, they can offer substantial rewards if successful, such as increasing your revenue by offering new services to your existing markets or entering new market segments (32). However, to succeed, these opportunities need a strong foundation, including a clear strategic vision, investment, and a culture of innovation. These factors make such opportunities far more difficult for competitors to mimic. By the time competitors recognize the outcome of your strategic innovation, the gap between your capabilities and theirs might be too wide to copy.

One challenge with radical business opportunities is that they are more difficult to find than incremental ones. When we surveyed about sixty business executives and asked them to share potential AI opportunities for their company, we captured over 700 ideas and opportunities. The number of ideas they shared was impressive; however, after analyzing the survey, we noticed that about 95% of the ideas were incremental business opportunities addressing existing problems and processes rather than ideas that could create new opportunities for their companies.

Why are radical business opportunities so hard to find?

Challenging existing assumptions and imagining entirely new ways of doing business can be difficult, especially in established companies. Radical business opportunities require a deep understanding of unmet customer needs, market trends, and potential disruptions. Unlike incremental business opportunities, which often are based on existing knowledge and practices, radical opportunities require creativity, stepping out of comfort zones, and the willingness to think beyond usual approaches. As Rob Sinclair, the Corporate Technology Lead at Wade Trim, emphasized, "People get stuck in their ways and old habits, making it extremely difficult to think out of the box and find radical AI opportunities."

How to find radical business opportunities

Finding radical business opportunities requires a paradigm shift in perspective. You need to look beyond the usual and explore novel and bold ideas. To assist you, in the following sections we share examples of strategic innovation outside of (or adjacent to) the AEC industry and a couple of questions to help you find new opportunities, and then provide some examples or ideas for radical business opportunities within the AEC industry.

Enter new market segments

Like the farmers and railroad example we shared in Chapter 1, strategic innovations can enable you to enter a new market segment. Komatsu, a Japanese multinational corporation founded in 1917 that primarily focused on copper mining, is a great example of a strategic innovator entering a new market. As copper became more scarce, Komatsu recognized the need to adapt. They invested in the manufacturing of heavy machinery and developing construction and mining equipment. By doing so, Komatsu eventually established itself globally. Now it is known primarily for manufacturing construction, mining, and military equipment.

Fast-forward to the 21st century, when Komatsu saw new opportunities in data and AI: every machine generates data, so the company leveraged this to develop a "Smart Construction" platform, which connects data related to on-site workers and machinery to increase job site safety and productivity (33). With this radical shift into AI and digital solutions, Komatsu expanded its services and entered a new market segment beyond traditional machinery manufacturing, making it possible to position itself as a leader in the future of construction—at least in Japan.

Like Komatsu, you can leverage your data and AI technologies to enter new markets and create the future. This approach is particularly relevant when seeking growth within existing or adjacent markets or identifying underserved segments. With this context in mind, what markets is your company serving today? Within this market or adjacent markets, is there any underserved market segment (i.e. a segment that needs your services, but cannot afford them)?

Let's explore an example to make things more tangible.

Imagine your company specializes in creating models that predict how rainwater will affect wastewater systems. You've been successful in winning large projects with medium to large municipalities in the U.S., helping them manage stormwater and prevent flooding. You deliver high-quality work and, in return, charge more based on your qualifications. In other words, you provide premium services.

One opportunity for you could be tapping into underserved markets with limited budgets. For example, a significant number of municipalities in the U.S. are small—nearly half of U.S. cities have fewer than 1,000 residents, and cities with more than 50,000 residents account for only four percent of all municipalities (34). These small municipalities often have limited budgets for designing and modeling infrastructure such as sewerage systems. Your company can utilize AI to reduce operational costs and offer services to these small municipalities at a lower price while still making the same or higher profits.

So, the question that you should discuss with your team is:

———————————————————

What underserved market segments might you serve with increased efficiency by using AI? How different are the needs of this market segment compared to the market you're serving today?

———————————————————

Offer new services

Earlier, we talked about John Deere's evolution through each wave of innovation over nearly two centuries. So far, their approach to the sixth wave of innovation, AI, has been to transition toward a software-as-a-service (SaaS) business model. For example, their new See & Spray AI-powered technology helps farmers kill weeds without wasting herbicides (35). By developing this technology, John Deere charges not only for the hardware and its installation but also an additional fee per acre treated. Offering See & Spray as a new service helps farmers save on the cost of wasted herbicides and earn value.

John Deere's use of AI technologies to provide new services to its customers shows how traditional industries can innovate.

But how might you find new services and offerings for your company? A practical way to find new service offerings is to ask yourself:

"If you could use AI to design and/or build ten times faster and better, how would you use the extra time and resources?"

This question often leads to insightful and creative discussions. Some executives say that if they could design or build ten times faster, they would use the extra time gained to improve the quality of their work, ensuring it exceeds clients' expectations. Others say they'd use a portion of the time saved to improve quality, and the rest to work on new, radically different ventures, including developing and offering entirely new services.

This approach not only encourages thinking beyond current challenges and limitations but also prompts new ideas and creativity, helping to explore untapped opportunities. By imagining the possibilities that come with increased capacity and resources, you can identify potential new services that were not previously possible or achievable.

Here are some insights shared by forward-looking AEC executives in response to the question above (*If you could use AI to design and/or build ten times faster and better, how would you use the extra time and resources?*):

As a design and engineering company, we would make a deal with the owners to maintain the buildings so that we can collect data on the actual performance of the buildings. Based on all the projects we design in the U.S., over time, we have building performance data at the city and country levels and sell those insights back to future clients.

The CEO of a construction company emphasized:

We have had some repeat clients over the past couple of decades. We have all the data from their projects in our infrastructure. When they have a question about the current project, we can use AI to analyze similar past projects and answer their questions based on the actual data and facts from their own projects. It is as if we know them better than they know themselves. This is how we build trust with them and keep them as our loyal customers.

Another example is from Gilles Caussade, the former CEO of ConXtech, an engineering, fabrication, and construction company. ConXtech developed a unique technology for fabricating special beams and columns that, like kit-of-parts, connect and make on-site assembly much faster. Despite the efficiency of this technology, ConXtech faced a major issue: the value of the technology was not immediately clear to clients, especially when compared to more conventional steel or timber structures.

To win projects, ConXtech's engineering and estimating teams had to spend lots of time designing building structures and demonstrating to potential clients how their system could save time and reduce cost. This process was extremely inefficient, costing them millions of dollars each year. To address this, ConXtech decided to implement an AI-powered configurator tool, which allowed its potential clients to directly compare the cost and assembly time of ConXtech's structural system with traditional steel options. This made it easier for clients to see the value in choosing ConXtech.

Gilles also planned to give this tool to the engineers working with their repeat clients, enabling them to compare structural systems on their own. This strategy not only could potentially help to increase revenue, but also could reduce inefficiencies that were costing the company millions of dollars.

ConXtech used an AI-powered tool to demonstrate the value of their services to their clients and turn a costly and inefficient sales process into a more effective and profitable strategy. So, the question that you should discuss with your team is:

_____

How do you leverage data and AI to offer
new services to your current or future clients?

_____

Create new revenue models

What would you do if a potential client said, "We need your services, but we don't have the budget to cover the full cost?" Do you walk away, or do you think about how to monetize your services differently?

_____

What would you do if your services were free for your clients?

_____

Most municipalities have limited budgets for managing and maintaining the infrastructure in their cities, but they need to build and maintain new facilities all the time. Imagine you have a potential municipality client who wants to build 1,000 new bus stations but doesn't have enough budget to do it. What would you tell them? To come back later, go away, or something else?

Let's see how Jean-Claude (JC) Decaux, a visionary French businessman who founded the advertising company JCDecaux in 1964, solved this problem. JCDecaux was initially a street furniture company that evolved into a global leader in outdoor advertising. The company revolutionized its business model by offering cities a deal that was hard to refuse: they would design, manufacture, and install bus stations for free. Yes, free! This deal was both a game-changer and a no-brainer offer. The catch? In return, JCDecaux asked for the right to use the bus stations they built for advertising. They monetized bus stations by displaying ads and therefore created a steady stream of revenue. With a global presence in over 4,000 cities in more than eighty countries, JCDecaux had a market cap of $4.5 billion in 2024. So, the question is: using the power of AI, what services can you offer that provide recurring revenue rather than rely on one-time charges? How big would that revenue stream be?

How might you monetize your services and offerings in a way your competitors have not imagined?

AEC companies often follow pricing models defined in contracting types, such as lump sum, cost plus, time and material, and guaranteed maximum price. In addition to these common pricing structures, AI could open up new revenue models, such as subscriptions to your services or licensing your data or services.

Although the data licensing and subscription business model may not work for every sector in the AEC industry, it could work in some market segments, such as infrastructure maintenance, building management, and even specific building products such as HVAC systems or elevators.

An example of successfully changing a traditional business model to subscription during the Internet disruption wave is Hilti in the early 2000s. Hilti shifted its outdated business model from product to service, a.k.a. Tools as a Service (TaaS), which provided its customers with a set of tools for a fixed period in exchange for a monthly fee. To implement TaaS, Hilti had to restructure its sales process, build fleet management infrastructure, integrate new contractual obligations into its IT systems, and even handle increased operational costs. Despite initial resistance, Hilti's executive team allowed enough time, over fifteen years, for gradual change management and scaled up step-by-step (36). Dr. Christoph, the former CEO of Hilti, credited this shift with helping them withstand the Global Financial Crisis of 2008.

Adopting a new revenue model not only provides your company with a steady revenue stream but also strengthens client relationships, making you the go-to place when assistance is needed. The value your clients receive over time far exceeds the subscription fee, making it a no-brainer offering.

What new revenue models can you create based on the value you provide to your clients?

As we consider the broader picture of business growth, we have discussed two types of business opportunities so far: incremental and radical. Which one should you pursue?

To disrupt the status quo by strategic innovation in the age of AI, you need to successfully implement a balanced mix of both incremental and radical business opportunities. A good rule of thumb is to allocate 80% of your efforts and investment to incremental business opportunities and 20% to radical ones. Doing so enables your company to continuously improve and maintain competitiveness in the short term while positioning itself for transformative growth and market leadership in the long term. By diversifying your innovation efforts in this way, you can better manage risk and remain adaptable and resilient in an ever-changing business environment.

Racing to the top with AI

At the beginning of this chapter, we introduced the business model as a description of how a business generates value and captures revenue in proportion to the value it creates.

For traditional companies, adding more value means adding more people to the projects to deliver the value. For instance, assume you have a traditional insurance company in the AEC industry, and you want to add value to your services by providing insurance products tailored to your clients. To do that, you need to allocate more people to research clients, measure their risk profiles, and analyze data to tailor products to customers' needs. Even though the added value could potentially increase your revenue, the increased operational cost may prevent your profit from increasing. Therefore, for many traditional companies, adding more value without increasing operational costs leads to cutting corners and reducing the quality of work. As Brad Miller, the CIO of Moderna, highlights (37), "Where traditional operating models require large workforces as business scales, real-time AI companies scale the value of their people."

How can you differentiate yourself in the market by adding more value to your client's work while reducing the operating cost?

This simultaneous pursuit of creating more value for your clients while reducing the cost of creating the value is called "value innovation," which is the central concept in W. Chan Kim and Renée Mauborgne's book, *Blue Ocean Strategy*. The book explains that you don't have to choose between increasing value or reducing operating costs. Innovative companies can have both.

Let's take AXA, a global insurance company, as an example. AXA decided to offer strategic and customized insurance products to their clients without increasing their operational costs. AXA uses AI to automate complex tasks like flood risk assessment and document processing for natural disasters. They also use AI to process large amounts of data, such as weather patterns and topography, to more precisely evaluate flood risks and offer strategic and customized insurance products to their clients (38). By using AI, they reduced their operational costs while still delivering high-value products to their clients.

What incremental and radical business opportunities can help you increase the value to your clients while reducing operating costs? And how can AI help you achieve that?

So, how to race to the top instead of the bottom? Like AXA, you need to identify opportunities to add significant value to your services while simultaneously reducing the operational costs of delivering that value. If you price your services based on a percentage of the value delivered to your clients, this approach can lead to increased revenue with lower operational costs, resulting in a higher profit margin.

CONCLUSION

In this chapter, we highlighted the critical role of both incremental and radical business opportunities in shaping your company's future and becoming a strategic innovator. While incremental opportunities focus on steady improvement in the existing services and processes, radical opportunities drive transformative change, including opening up new market segments, providing new offerings and services, and creating new revenue models. By understanding and strategically balancing these two types of business opportunities, you can position yourself for long-term success and deliver tremendous value to your clients. As we have seen, leveraging AI can redefine your business models and increase value delivery while reducing operational costs.

In the following chapter, we will discuss how AI impacts your operating models and how to prepare your data, processes, and people to successfully implement these radical opportunities.

WADE TRIM'S JOURNEY IN CREATING AN INNOVATION STRATEGY

Wade Trim, a civil engineering firm headquartered in Detroit, Michigan, was established in 1926 as a small local business focused on designing roads and sewers. The company has since grown into a national powerhouse with over 800 employees and twenty-two offices across the U.S. As a 100% employee-owned company, Wade Trim benefits from a unique ownership culture that fosters its work approach, collaborative culture, and quality final projects. The firm is regularly recognized among the top 200 design and top 500 environmental firms by ENR.

We met Andrew (Andy) McCune in a virtual session, during which we delivered a talk about AI opportunities in the AEC industry. After the session, Andy consulted with his executive team and started Wade Trim's AI journey. The following section summarizes Wade Trim's journey into creating an AI strategy from Andy's point of view.

In recent years, advanced technology has become a cornerstone of our approach to innovation. With the help of Tim O'Rourke, our Chief Technology Officer, we started our Office of Applied Technology (OAT) to proactively seek out emerging technologies to test, nurture, and develop ideas to stay technically proficient and competitive in a fast-changing innovation landscape.

Noticing how other firms in the AEC industry were leveraging AI to improve efficiency, we recognized the need to explore AI and the opportunities it provides to our business to avoid falling behind. That said, we wanted to approach this cautiously because we knew it would be risky to rush it and jump into the ocean without understanding what AI really is.

To kickstart our AI journey, we hosted an AI summit, bringing together thirty key executives and members of our newly formed AI taskforce for a full-day workshop. Initially, some executives were concerned about dedicating so much time and involving many employees, but these concerns quickly faded. The participants expressed that AI education was valuable and inspiring and even generated a sense of excitement and enthusiasm. It also enabled a culture of innovation, setting the stage for the leadership team to craft their AI strategy.

Following the summit, we engaged the task force to identify innovative business opportunities. Through a structured framework, the task force initially identified over twenty potential opportunities, including incremental and radical. These were eventually narrowed down to three key opportunities best suited for the company's growth and establishing market differentiation by leveraging AI. For instance, one of the radical opportunities identified by our task force was an AI-powered wastewater management tool, which, when built in-house, we can utilize to offer our services to an entirely new market segment of municipalities and industrial clients beyond our current market and client base, significantly increasing our revenue and market share.

Our AI strategy gave us clarity on who we should hire to stay competitive in the age of AI. In addition to our regular hires and AEC subject matter experts, we've been hiring AI experts as interns and full-time employees. We see a future when the AI and technologies we build in-house and our data become our main differentiators in the market.

CHAPTER 6

Operating Model as an Enabler

Karl was the CEO of a major technology company in the AEC industry. For years, the company sold perpetual software licenses (products delivered via CDs or downloads) that customers could own indefinitely. It was basically a one-time sale unless customers wanted to upgrade to a newer version by purchasing maintenance licenses. The business was booming, with the company dominating the market.

However, as cloud technology became increasingly available, Karl recognized both the opportunities and the threats it presented. On one hand, the opportunity to leverage cloud computing was increasing the company's agility, reducing software piracy, and improving customer experience. On the other hand, the threat of ignoring this technology shift could mean losing their hard-earned market share to more forward-looking competitors. Technically, the risks of not investing were high.

In a board meeting, Karl shared his idea of transitioning their products to the cloud. While most board members were interested and curious to hear more, one of them had many questions and concerns:

- "How many customers have actually asked for this? Are they even interested?"
- "Why not wait a few years and see how the market changes? I think you're acting too fast."
- "This is a huge investment for the company; how can you guarantee its ROI?"

Karl listened calmly before responding, "Our customers may not yet realize the full potential of cloud technology and still don't even know what cloud is. It's our responsibility to educate and guide them through this transformation. This isn't just about investing in a technological upgrade; it's a strategic move to secure our company's future in an increasingly cloud-driven market."

After several meetings and conversations, Karl finally obtained the board's approval and executive buy-in to start the transformation. Karl and his leadership team knew this was a strategic change for the company and would not be an easy and straightforward journey.

Their biggest challenge was a lack of skills or expertise in developing cloud software. The company had over 3,000 software engineers skilled in desktop development, but not one was an expert in cloud software. Besides that, this transformation required a fundamental change—from selling perpetual licenses to offering subscriptions. This change had to go beyond the financial department and required an entirely new business model.

Under the new subscription model, the company shifted its focus from making a one-time sale to providing continuous value and support. So, they created a dedicated customer support division to ensure ongoing customer satisfaction and retention. The sales division also had to reinvent its entire sales approach and educate clients on the long-term benefits of cloud services. This task was not easy, as many customers initially resisted the change.

After several years of hard work, including cutting some internal operating costs to meet this vision, the company successfully transitioned to the cloud. Their clients eventually saw the value in this shift. The company could also expand its offerings, enter new markets, and adopt a more flexible pricing model.

Like Karl's story, oftentimes, leveraging new technology disruptors like AI is not merely a technological transition. It's a transformational initiative that impacts various aspects of the business and its operations. In their book, *Competing in the Age of AI*, Karim Lakhani and Marco Lansiti talk about this transformation as being more than a technology initiative: "The challenge is restructuring how the firm works and changing the way it gathers and uses data, reacts to information, makes operating designs and executes operating tasks."

Back to Karl... Imagine he had given this initiative to his CTO. Perhaps the CTO would have tried it on a small scale, but the project would have failed because of the lack of resources (e.g. cloud software engineers), other competing priorities, and a lack of known ROI. Moreover, since this transition was not just technical, the CTO would likely have struggled to get buy-in and commitment from other departments. After all, without an initial clear customer demand, the CTO would have found it challenging to justify prioritizing this initiative.

As mentioned in earlier chapters, one of the problems we see in the AEC industry is that many CEOs and executives view AI as a technology initiative and dedicate it to their technology team.

The point is that strategic innovations require deep operational planning, and they're not quick fixes or short-term gains. Therefore, in this chapter, we will discuss how you should rethink your operating model—more specifically, your workflows and processes—to capture better data for your AI initiatives (Figure 6.1). Then, we'll explain the structure of your AI team and the network effects of AI within your organization, helping you understand the holistic view necessary for a successful AI transition.

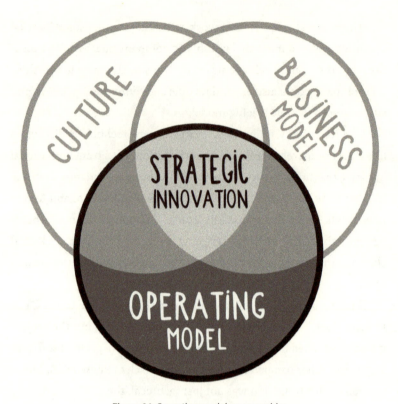

Figure 6.1: Operating model as an enabler

Treat your data as an asset

We often hear from CEOs and AEC executives that they are sitting on twenty, thirty, or forty years of data, depending on the age of their companies. Sounds valuable, right? You may have heard the messages in the media about how "data is the new oil" or "data is gold."

But in its current state, you should consider your data as "garbage." Some are recyclable and useful (e.g. some of your financial data), some are compost with a limited lifespan (e.g. data related to the cost of materials), and some are trash (e.g. inconsistent and bad data). If you don't put the right segregation processes in place, you'll end up with a mixture of different types of garbage in one place, which makes it extremely difficult to get any value out of it. You won't know what is what! Now, imagine you have collected a mixture of data over many years without proper tagging and metadata. Some are expired because codes have changed, and some are not. So, untangling this mess would be a very time-consuming and expensive endeavor.

It's not about how old your archived data is;
it's about when you began treating it as an asset.

Sam Ishak, CIO of Langan, an engineering and environmental consulting company and in the top fifty on ENR's 2022 500 Design Firms list, shared his perspective on data in AEC: "Our AEC industry generates vast amounts of intelligent data that remain trapped in unintelligent repositories. Partnering with AI now will allow us to harness and preserve that knowledge, not only for today but for the generations to come!"

But how do you treat your data as an asset?

The answer depends on who you ask. The CIOs, CTOs, or IT directors in AEC companies often want to develop their own data lake or data warehouse, so they tell their CEOs that "we need to have high-quality data before talking about our AI initiative, so we should invest in centralizing our data or creating a data lake."

The drawback of a data lake is that once it is developed, it stores massive amounts of data from various sources in different formats and of varying quality, often without proper policies and guidelines. Moreover, what often happens is that everyone just dumps all the data there so they don't lose it, and eventually the data lake turns into a data swamp.

We have also seen companies spend millions of dollars creating data lakes without even answering the simple question, "What is it for?" Collecting data for the sake of it is not the proper approach.

Also, accessing data in a data lake is often challenging. We were working on an AI initiative for a major tech company and needed to access user data, such as roles, levels of experience, and project types, stored in the data lake. However, the legal team had concerns about data privacy and access because not everyone was authorized to access all data points. This blocked us from using the data we needed and slowed our progress. In the end, the legal team banned using the data.

To be clear, we are not against data lakes, but we do not recommend creating one as the first step in turning your data into assets.

So, what is the first step?

Instead of starting with technology like data lakes, we recommend starting with your processes and people. Let us discuss this in the next section.

Standardize workflows

Remember the innovation waves we shared in Chapter 1? One of the major technological innovations in the third wave was the creation of Henry Ford's Model T in 1908. This innovation was so profound that people often associate the invention of the car with Ford. However, the reality is that cars were invented a long time before Henry Ford built the first Model T. For instance, in Germany in 1885, Carl Benz built the first modern car suitable for daily use (39). So, what was Henry Ford's real innovation?

The short answer is: his process.

Back then, automobiles were mostly handcrafted from the ground up by skilled artisans. The parts were often made specifically for each vehicle and machined by hand. In those shops, vehicles stayed in one

OPERATING MODEL AS AN ENABLER

place, and craftspeople had to move around with their parts and tools to build them. The workspaces were small, and the process was ad hoc, time-consuming, and disorganized.

Ford introduced the moving assembly line, which allowed vehicles to move down the line instead of having workers moving around them. While the new process enabled interchangeable parts, allowed faster assembly, and subdivided labor, it also required fundamental synchronization and standardization (40). Ford's assembly line changed how cars were produced and transformed automotive manufacturing processes.

How does this relate to your situation?

Your organization likely has established processes for delivering projects to clients. For instance, when you win a project, project managers take charge by planning the details and requesting resources. Once those resources are allocated, the project moves forward, with everyone focused on delivering all requirements outlined in the contracts.

However, one of the challenges we often see is that these processes are not standardized and documented. Every division in the company has its own way of completing tasks, using different tools and processes. These methods often depend on or are determined by the project managers and people working on the individual projects. In other words, if you have twenty teams working on projects and want to document the process of delivering these projects to clients from start to finish, you'll end up with at least twenty different process maps.

The second challenge is that almost none of the companies in the AEC industry have standard procedures for how project data should be managed and delivered within their own company. Project team members work hard to complete projects for clients, and once a project is finished, they should move on to the next one. But what happens to all the data generated during the project? Often, it ends up scattered across different platforms—stored on employees' desktops, OneDrive, or other cloud solutions. Up to thirty percent of initial data created during the design and construction phases is lost by project closeout

(41). This underscores why your data does not become an asset until you start treating it as one.

Why does this matter?

Allowing everyone to perform tasks in their own way leads to inconsistency and poor data quality, or even data loss, risking the success of your AI initiatives. As Bill Gates pointed out, "The first rule of any technology used in a business is that automation applied to an efficient operation will magnify the efficiency. The second is that automation applied to an inefficient operation will magnify the inefficiency." If your current processes resemble the disorganized car shops of the pre-assembly line era, then it's time to standardize. Like Ford's assembly line revolutionized manufacturing, you need to standardize your processes and establish project data delivery procedures in your organization to succeed in your AI initiatives.

As Shane Danley, director of technology at Parkhill, an architectural engineering company based in Texas, said, "Without standardized workflows, integrating AI into AEC companies becomes a fragmented and inefficient process. By standardizing our workflows, we can ensure data consistency and enable faster adoption of new AI-enabled systems to maximize the potential of the technology through a unified and scalable approach."

Automating broken processes will not get you far, but if you rethink them before automating, you'll see real progress.

But how can you actually implement standardized processes within your organization?

Examples of process standardization

Let's take a look at examples of how process standardization was achieved in two different companies.

We used to work in an organization with offices in several locations. As project managers, we had full autonomy over how to run projects, as long as our directors were okay with that or as long as the work had been completed. The company directors had their own style and process for managing projects. Some were hands-on and asked for daily updates; others were hands-off and occasionally asked us how things went. How we documented projects and chose platforms for sharing files was also up to us. Some managers used Microsoft Teams, and others used OneDrive or Google Drive to share files.

The challenge with this approach was that leadership had no visibility into how many projects our department was handling at any given time, what their current status was, or what the potential blockers were. To get this information, they had to email us and ask directly. As time passed, with teams growing and more projects being taken on, the situation got worse. The company had various restructurings, and many project managers moved to other departments, retired, or left the company. As a result, no one knew what was happening in our department until the head of the department changed.

When Sara, the new head of the department, came on board, she hired Jack as a program manager and worked with him to create and implement a new process for how everyone in the department should work and what workflows should be.

In one of the all-hands meetings, Sara announced a new initiative. She shared, "From now on, all project managers must use Wiki, a cloud-based collaboration platform, to create projects, update status and report progress, share challenges, and share lessons learned and project deliveries." We were shocked! Someone asked, "Is this in addition to the main task of delivering the work to clients?! We're too busy. We have no time for this!"

Sara listened to the employees' concerns. To support their time management and maintain consistency, she asked Jack to create several templates for onboarding new projects, status updates, and project closures. Every project manager had to use those templates. Every month, Sara began her all-hands by showing a report on all the projects from the Wiki, and she publicly recognized some of the project managers to motivate everyone to utilize the Wiki.

Within a few months, all project details were accessible on our Wiki platform. Eventually, we concluded that even though adding details to the Wiki was an additional task, this initiative significantly reduced the time spent replying to individual emails requesting information about the various aspects of a project. It saved us hours of searching for information, prevented duplicated efforts, and made it easy to track lessons learned.

The leadership team now had a dashboard of the portfolio of projects. It was easy to find out which projects were in progress, who was involved, and what the status of the project was. This new process brought visibility to our efforts in the department and helped identify high and low performers.

You need to establish a standard process for creating, documenting, and delivering projects to your company.

Sara was extremely patient with the change she introduced to the organization and provided essential training for us. But Sara's story shows just one way of standardizing processes. Let's look at another case where a different approach to standardization also led to a significant gain in how projects were managed and executed.

Back in 2002, Amazon had a critical business challenge in operation. At that time, Amazon was rapidly expanding its business, and its internal software systems were becoming very difficult to manage, scale, and integrate. Different teams within Amazon were building systems in silos, leading to a lack of standardization and communication between services. As a result, integration issues began to surface across the organization. To address this, the new mandate required that all Amazon teams design their systems so that they communicate with each other through service interfaces, specifically APIs (Application Programming Interfaces). This is how Jeff Bezos approached the process of standardization:

FROM: Jeff Bezos

TO: All Development

SUBJECT: Bezos Mandate

All teams will henceforth expose their data and functionality through service interfaces. Teams must communicate with each other through these interfaces. There will be no other form of inter-process communication allowed: no direct linking, no direct reads of another team's data store, no shared-memory model, no back-doors whatsoever. The only communication allowed is via service interface calls over the network.

It doesn't matter what technology they use. All service interfaces, without exception, must be designed from the ground up to be externalizable. That is to say, the team must plan and design to be able to expose the interface to developers in the outside world. No exceptions. Anyone who doesn't do this will be fired. Thank you; have a nice day!

Jeff Bezos

We're not suggesting you adopt Jeff Bezos' approach and attitude to standardize your process! However, his API mandate is an excellent example of how standardized processes and governance can drive innovation and success. His approach led to the creation of AWS, which is now a billion-dollar business for Amazon.

To sum up, as a strategic innovator in the AEC industry, you should have standard processes for workflows in your organization. Without them, you won't have the high-quality data needed for your AI opportunities, as inconsistent processes can lead to data gaps and inaccuracies in your AI systems. In addition, you should develop standard processes for collecting and delivering project data to your organization. Together, these practices lay the foundation for implementing AI opportunities and their success.

Still, even with standardized processes in place, a common question is: Who is responsible for the quality and consistency of data? In the next section, we will look at the roles responsible for managing data—because without clear ownership, even the best plans can fall apart.

After all, who is responsible?

At one of our AI summits, Victoria, the head of IT, and Jeff, the head of operations, sat next to us during the networking session. Their conversation was about a common issue in AEC companies: whose responsibility is to manage data in AI initiatives?

Victoria shared, "If we want to leverage AI in the company, we need to be more intentional about our data."

Jeff replied, "I agree, but isn't that an IT responsibility? Data management falls under your domain, doesn't it?"

Victoria responded, "No, Jeff. This is operations data. The IT team isn't responsible for data. We can build the infrastructure and data pipelines, but the ownership of the data itself, particularly operational data, stays with your department. This is not just an IT task."

With a glass of wine in hand, Jeff smiled slightly and said, "But we're already stretched with our current projects. Adding data management for AI initiatives feels like an additional burden. I'm not even sure this AI thing will work for us when we're already so busy..." Jeff obviously was not comfortable with where the conversation was heading.

This wasn't the first time we had heard this debate about who is responsible for data. In almost all conversations with CIOs, COOs, and CTOs, they debate data ownership for AI initiatives. There's often confusion about who should drive these efforts and who's responsible for governing them.

To succeed in your AI initiatives, everyone should recognize that data is a shared responsibility beyond IT or operations. While IT is responsible for building and maintaining the infrastructure and data pipelines, the quality and accuracy of the data depend on the teams that generate and use it, like operations. But it doesn't stop there; every employee should understand and take ownership of their part in managing (e.g. producing and delivering) data so the company can make the most of its data and achieve its business goals. Salla Eckhardt, CEO of SARA Oy, a Finnish construction project management company, said, "If people are generating or documenting data, they need to stand behind it. I hope that, over time, everyone will contribute to building a reliable digital record, one bit at a time."

Data is everyone's responsibility: owned, managed, and delivered by all to drive success.

Recognizing roles and responsibilities is essential, but becoming a strategic innovator also requires implementing your AI initiatives effectively. To do so, you need to find the right AI team structure, which we will discuss further in the next section.

Structure your AI team

When it comes to implementing your AI initiatives, you need a combination of AI management and AI development teams. You can choose to hire an in-house AI development team or outsource it, but you still need an internal management team in charge to drive and track initiatives.

For your internal AI management team, you may consider some functional roles such as Chief Data Officer (CDO), Chief AI Officer, and AI Risk Officer. Your Chief Data Officer is responsible for managing data as a strategic asset. They oversee data governance, quality, and privacy across your organization and play a critical role in maximizing the value of AI initiatives. Meanwhile, your Chief AI Officer focuses on strategically implementing AI technologies and aligning AI projects with business objectives. Your AI Risk Officer, on the other hand, is responsible for identifying, assessing, and mitigating risks associated with AI adoption and can help set policies for assessing whether the AI solutions your employees adopt are secure and compliant with your company's value culture and regulatory standards.

As you think about hiring these roles, you should note that one size doesn't fit all. Hiring new roles depends on various factors, such as the size of your company, business vision, budget, and needs. If your company is small and has limited resources, one person may fill all these roles. Regardless of your approach, you need to have the right leadership and expertise to successfully integrate AI into your organizational DNA.

But how about your AI development team?

Depending on your strategy and resources, you must decide whether to create AI development tools in-house or hire an external team. If you decide to hire in-house, you should consider some roles such as AI engineers, data scientists, and business analysts. You should structure your AI development team in a way that enables agility and rapid experimentation, which are essential for driving strategic innovation. There are multiple ways to do so, but the primary structures are centralized AI, decentralized AI, and hybrid AI teams:

- **Centralized AI Team Structure.** A centralized AI structure could be ideal for smaller companies or those who have just started their AI journey. That said, this does not mean that this structure is solely for small companies; a lot of large companies also use this structure. This structure consists of a dedicated AI team that operates as a central hub and serves your entire organization (see Figure 6.2). The team will work with all departments and divisions within your organization and help anyone with AI needs. A centralized model enables consistency across your organization and allows you to consolidate AI expertise in a single team, which can enhance resource management efficiency and align with the company's overall strategy in the short term. However, as AI ramps up across your organization, it might be difficult for the AI team to meet demands unless your organization provides enough resources to the AI team. For example, Microsoft has adopted a centralized approach for both its AI and broader operational frameworks to drive company-wide transformation. Through its centralized AI Center of Excellence (AI COE), Microsoft consolidates resources and standardizes practices across departments to prioritize business outcomes over individual team budgets. Their aim is to enable faster decision-making and better resource alignment across the company's diverse functions and products (42).

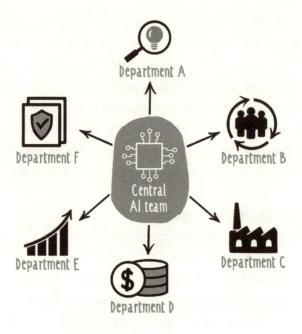

Figure 6.2: Centralized AI team structure

- **Decentralized AI Team Structure.** A decentralized AI structure may be a better choice for larger or more diverse companies, where different divisions or departments have unique needs that require specialized AI solutions. In this model, AI teams are embedded within each business unit or department and work closely with the specific teams they serve (see Figure 6.3). This model allows customized AI solutions tailored to the distinct objectives of each business unit or department, which promotes agility and closer alignment with the goals of each division. However, having separate AI teams in every unit can lead to inconsistency across your organization and might result in duplicated efforts if not managed and governed appropriately. You might be able to mitigate this drawback by maintaining a level of horizontal communication among the AI teams.

Figure 6.3: Decentralized AI team structure

An example of a company with a decentralized AI team is Meta, which embeds its AI expertise directly within each product team. Jerome Pesenti, Meta's former VP of AI, highlighted this shift as a way to make AI capabilities more accessible and better aligned with each product team's specific needs (43).

- **Hybrid AI Team Structure.** A hybrid AI structure is more appropriate for organizations that need centralized oversight combined with the flexibility to address specific departmental needs (See Figure 6.4). In this model, a central AI team is responsible for setting overarching strategies and governance standards, while decentralized teams embedded within different business units focus on their AI needs. This model allows your organization to leverage the central team's ability to standardize and streamline AI initiatives while also enabling decentralized teams to meet the unique needs of each business unit or department. Such a model promotes both strategic alignment with your company's overall business goals and operational responsiveness to department-specific requirements. The drawbacks of this model lie in the cost of hiring and retaining AI talent and ensuring communication and coordination between the central and decentralized teams to prevent conflicts and duplicated efforts.

An example of a company that uses a hybrid AI team structure is Uber. At its core, Uber's AI platform, Michelangelo, centralizes and standardizes AI efforts and development. This foundation is complemented by specialized AI teams embedded within product groups like Uber Rides and Uber Eats, allowing each team to meet specific needs while maintaining alignment with company-wide AI initiatives. This hybrid model enables product teams to innovate autonomously while staying connected to Uber's strategic goals (44).

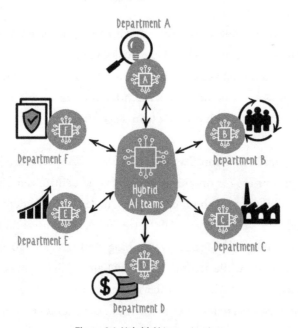

Figure 6.4: Hybrid AI team structure

After learning about the various AI team structures, you might wonder which one you should choose. The choice between a centralized, decentralized, or hybrid AI structure depends on where you are in your AI journey, your company's AI and data maturity level, your budgets, your organization's skill sets and talent, your business objectives, and your success metrics. Ultimately, selecting the optimal AI team structure is a dynamic decision you should revisit and refine as your organization evolves. Most likely, choosing a centralized AI structure could be a good starting point, as many AEC companies are small and still at the beginning of their AI journey.

But what if you want to outsource your AI development project by working with an AI tech provider? This is a common approach for companies that don't want to hire in-house. Based on our experience, there is one caveat with working with an AI tech provider, which, if not managed properly, could cause your company to lose hundreds of thousands of dollars.

Before explaining the caveat, you should know that each AI tech provider has strengths in certain areas. For instance, some companies specialize in computer vision and have many computer vision experts who can analyze photos. Others have many data scientists who can analyze time-series data (e.g., sensor data), while others may have mathematicians who are skilled in solving optimization problems. Because of their expertise, they tend to approach problems using the methods they know best or the techniques they are most familiar with.

Now when you approach these experts to develop an AI solution for the business opportunity you defined, they may, intentionally or unintentionally, try to solve it using methods that align with their expertise. However, this approach might not necessarily be the best solution. For instance, say you want to monitor how people move around a construction site. A computer vision consultant will most likely approach this by analyzing photos taken on site (from the CCTV cameras). In contrast, consultants with data scientists may opt to place sensors on workers' hard hats to track their data. But how would you know which approach has a higher probability of success for your project?

Besides choosing the right consultant, companies often start their AI development project without well-defined specifications. This situation is very similar to having vacant land and verbally explaining to your contractor what you imagine you want to build on it. In this case, the contractor wants to design and build the building for you. But rather than considering your best interests, the contractor may design and deliver the project based on resources available to them to increase their profit. You may not like the building you end up with.

A wise person will hire an architect to design their building, write the specifications, and act as their representative during the construction process. It's the same with an AI project—you need a knowledgeable consultant and partner to help you design the project, develop specifications, find the right AI tech provider, and guide you through the complexity of AI development. Missing this advisor could cost you hundreds of thousands of dollars and create avoidable failures.

To sum up, you can hire in-house AI staff or work with an AI tech provider or consultant to implement your AI opportunities. The choice depends on the size of your company, your budget, and your long-term vision. As you implement these AI opportunities, you need to understand how they impact your organization as a whole. In the next section, we'll discuss the cascading effects of AI opportunities on your entire organization.

Be aware of network effects

Imagine the different departments in your company as gears in a gearbox. A company that succeeds in its AI journey is one in which all departments (or gears) work together harmoniously and generate value for the business, clients, and stakeholders. If one department becomes extremely efficient by automating its processes without considering other departments, it may not necessarily benefit the company as a whole. In fact, it could have a negative impact. The "faster gear" may create resistance and tension from other gears or even break the entire gearbox.

For example, the marketing team might use an AI solution to write proposals to minimize repetitive and time-consuming work. With this AI solution in place, the marketing team can deliver more proposals faster and tailor them to clients' interests and needs. Despite the potential benefits the marketing team gains by using AI, this may not help the whole business because other divisions may not have enough capacity to take on more projects. What would be the benefits of submitting more proposals without having the workforce to do the extra work those proposals bring in?

After setting up an AI strategy and defining business opportunities, every department typically starts experimenting and learning about AI without considering the holistic impacts on other departments. Even though this approach makes your departments more agile, as you make progress in the long term, you should assess the impact of AI use on your entire company. Just because a team can use AI to do their job faster doesn't mean the whole process will be faster. The rest of the process may still take as much time as before or, even worse, the process could be slowed down by other teams.

Real progress happens when everyone moves forward in harmony and together.

It's not enough for one team to adopt high-end tech while others stick to the old ways. To ensure your entire organization evolves and grows together, you need to adopt a holistic approach that considers interdependencies between departments. Everyone should understand how changes in one part of your organization can cascade through the entire network.

In addition to considering network effects at a company level, real progress happens when the entire industry moves forward in harmony and together. Let's go back to the ConXtech example discussed earlier. Imagine if, with their specialized structural steel, they could finish installing a building structure in two weeks instead of three months. If this added value is not considered and coordinated, even if the team had completed their work faster and had the structural design ready ahead of schedule, the owner may not get the full value because the MEP (mechanical, electrical, and plumbing) contractor was still scheduled to install MEP components in the third month. The benefits of speeding up one part of the process are lost when the rest of the stakeholders continue to follow the convention. Similarly, AI's primary value for clients and stakeholders is when it is adopted across the entire network in the industry. As more and more companies across the industry adopt and integrate AI into their processes, the full benefits of AI become more evident.

As Scott Thomson, the VP of Technology at SSOE, mentioned:

Successful organizational leaders look at the entire system and workflow rather than a specific one. A chain is only as strong as its weakest link, and sometimes a stronger (more efficient) link needs to give up its strength (efficiency) to strengthen the entire chain. That's why we need to have a holistic view of how AI impacts the efficiency of our entire organization rather than just focusing on a specific use case without analyzing its network effect.

As you move toward broader AI adoption, you should track and measure your progress. In the next section, we will discuss how to set metrics and measure your progress.

Measure your progress

To measure your progress in your strategic innovation journey, you should define and set metrics to monitor its status. You must enforce and have these metrics tied to meaningful incentives. Depending on your company's culture and management style, you can use the 'stick' or 'carrot' strategy to enforce what people should do. According to Nectar, an employee recognition platform, a survey of 1,800 full-time employees in the U.S. showed that (45):

- 83.6% of employees feel recognition affects their motivation to succeed.
- 81.9% of employees link recognition of their contributions with how engaged they feel.
- 77.9% of employees would respond to more frequent recognition by being more productive.

As you can see, employees are more likely to be motivated and committed to achieving company goals when their efforts directly lead to rewards (e.g. bonuses, recognition, career advancement, or other tangible benefits).

Ensure your employees understand the rationale behind your initiatives and metrics and are committed. As Simon Sinek said, "People don't buy what you do; they buy why you do it." That is how you can build a culture of accountability and high performance.

So, what are some examples of these metrics?

Because no one has predefined metrics that could work for all companies, we encourage you to define metrics for your company based on your ultimate business goals. Here are some examples you can use to define specific metrics for your company.

Suppose you have a hub to capture innovation ideas. In that case, your team may track the number of ideas generated and measure whether they solve a problem or bring a new opportunity to your company. You may track the number of ideas relevant to incremental versus radical business opportunities. Moreover, you can assess the diversity of participants in ideation (e.g. is it just a group of people who consistently submit ideas or a diverse group of your employees?).

One of the steps in creating a culture of innovation discussed in Chapter 4 was shifting people's emotions with training and education. When you offer these sessions, you may track the number of employees actively participating or people who take AI courses on your learning and development (L&D) platform. Tracking employees' emotions about AI over time could be another way to check the pulse of your organization.

If you are bringing AI technology to your company, track its adoption rate and see if people push it back without any logical reasons. If so, this might be because they are still scared of losing their jobs and being replaced by AI. And don't forget to track your employees' satisfaction and retention rate to see if there's any positive cultural shift. Another metric to track might be the number of processes that have been standardized in the company or employees' and management's contribution to producing and delivering quality data.

These metrics, or similar ones, give you some idea of whether your investment delivers value or not. And always be patient when it comes to changing your people's mindset and culture. Any change takes time!

CONCLUSION

In this chapter, we discussed the critical roles of aligning data, processes, and organizations to succeed in your strategic innovation journey. We explored the benefits of process standardization, showing that significant improvements do not necessarily require huge investments. We also discussed how you can start by documenting existing workflows and adopting best practices to standardize your processes, reduce repetitive efforts, and increase data quality. You also learned that you should be aware of the network effects of your AI opportunities in the long term and the importance of setting metrics to measure progress.

MACKENZIE'S JOURNEY IN EXECUTING INNOVATION STRATEGY

For over sixty years, Mackenzie has been a leading force in the AEC industry, known for more than just its expertise in architecture, engineering, and planning. The company is a collaborative team of professionals dedicated to shaping the built environment while focusing on innovation.

We met Josh McDowell and Matt Butts, board members and principals at Mackenzie, before they started their AI journey. They were both passionate about the potential that AI could bring to their companies. The following section summarizes Mackenzie's journey into bringing AI into their operation from Matt's and Josh's point of view.

To maintain our competitive edge in the market, we began exploring how best to integrate AI into our operations. We recognized that AI was more than just a trend; it was essential for thriving in the future. Therefore, we decided to take a well-thought-through approach that involved the entire organization in this AI journey.

That said, adopting AI has its own challenges. Our employees were worried about how AI might impact or replace their roles (the "what-ifs" about job displacement were significant). Also, there was a widespread feeling that we lacked the technical knowledge to navigate this new technology. Despite these concerns, we moved forward, learning valuable lessons along the way.

One key realization for us was that data itself is not as valuable as one might think. We shifted our focus from just collecting massive amounts of data to refining processes that could turn data into assets. This more targeted approach emphasized the importance of data quality over quantity, allowing for a more strategic use of AI potential in our business.

We also recognized the importance of standardizing our workflows to adopt and leverage AI in our business. The more standardized and interconnected our processes become, the more effective and impactful our AI solutions will be. In other words, inconsistent processes could negatively impact AI's outcomes for our business and clients.

We see AI not as a tactical but a transformational initiative in which our board and leadership team play a key role. They actively set the tone for the entire company, guiding the standardization of AI strategy implementation and data management.

We formed an AI task force composed of employees from diverse backgrounds and experience levels to have a shared responsibility across the company. This inclusive approach created a sense of ownership and commitment among all employees.

We knew that we should be agile in this journey and consider it a continuous improvement process, so we adopted a data-driven approach to refine our AI strategy and measure our employees' engagement in moving this initiative forward.

Our journey to execute our AI strategy includes strategic planning, process standardization, and inclusive leadership. Aligning AI with our unique company culture and involving the entire organization, we successfully embraced AI and also positioned ourselves for future growth. Our strategic approach transformed challenges into opportunities, helping us stay ahead in the competitive AEC industry."

FINAL WORDS

We wrote this book to help CEOs, board members, and executives of AEC companies radically innovate and transform their organizations in the sixth wave of innovation (AI). In any innovation wave, you should disrupt your company's status quo and become a strategic innovator; otherwise, you risk being disrupted by others.

In Part 1 of this book, you explored the six waves of innovation and AI. With increased data, enhanced computing power, improved algorithms, and major capital investments, we're at the beginning of a massive AI disruption. You learned that every AEC company is now, in essence, a technology company, one that can leverage data and AI to develop solutions unique to your business. This creates an unprecedented opportunity for innovation within your organization.

In Part 2, you learned about the three pillars of strategic innovation. You learned how to create a culture of innovation in your company by leaning in, setting a budget and the right expectations for ROI, shifting your employees' emotions from fear to excitement by providing training and education, and celebrating failures as learning opportunities. Companies like SSOE lead the industry by fostering such a culture of innovation.

Next, we discussed the importance of inventing or reinventing business models in each wave of innovation. You learned that this involves creating new services for clients, expanding market segments, or creating a new revenue model. Companies like Wade Trim have found these business opportunities and are executing them to gain new market share.

Finally, you learned about redefining your operating models to treat your data as an asset. You learned that most data and AI-related problems in our industry are not technology-related problems; they are people-related problems. We shared that by standardizing your processes and setting the right governance and incentives, you will be able to collect consistent data across your organization. Companies like Mackenzie are standardizing their processes to get the most out of their digital assets—data.

By fostering a culture of innovation, actively pursuing new opportunities, and redefining your business and operating models, you transform your company into a strategic innovator that continuously disrupts the status quo. By leveraging AI and harnessing high-quality data across your organization, you deliver tremendous value through your services while reducing operating costs. Unlike conventional AEC companies, the industry's resource constraints do not affect you because you have focused on scaling the value of your talent through data and AI.

Due to your strong culture of innovation, you attract top talent in the industry while maintaining the highest employee retention rates. Your clients love the value of your services, which are unmatched by competitors, leading them to offer you more projects. This allows you to be selective about whom you work with. More and better projects generate more valuable data, which further empowers your AI ecosystem and delivers even greater value to your clients. This creates a virtuous cycle of continuous improvement and success.

We hope this book has given you the insights, tools, and inspiration to lead your organization through the sixth wave of innovation, AI. Whether you're a CEO, board member, or executive, you are responsible for driving change, disrupting the status quo, and ensuring your company not only survives but radically innovates.

Now, as you finish reading this book, remember that the innovation journey is not a sprint but a continuous process of learning and growing. Over time, the only way to retain your knowledge is by taking action. Edger Dale's "Cone of Experience" theory states that we'll remember

10% of what we read, 20% of what we hear, but 90% of what we do (46). So, what are you waiting for?

To help you achieve that, we have created a small companion workbook with all the information you need to succeed in leading your company and the AEC industry in the age of AI. The workbook contains additional reading lists and resources to help you in your AI journey. You can download the workbook at www.disruptit-book.com.

As you move forward, remember that innovation is as much about the journey as it is about the destination. It is a never-ending journey. Enjoy the ride, encourage those around you to do the same, and together, let's shape the future of the AEC industry.

We hope you will disrupt the status quo—within your company and across the industry—for the benefit of all!

CONNECT WITH US

Thanks for taking the time to read *Disrupt It*. It means a lot to us.

We hope this book has provided valuable insights and answered questions you might have had about operationalizing AI within your organization. If you have any questions or feedback or would like to share how this book has impacted your organization, we would love to hear from you. Let's keep the conversation going!

If you are looking for additional resources and offerings, we invite you to visit our website at www.yegatech.com. You may also like to consider the following:

- AEC Disruptors Circle: Join our AEC Disruptors Circle, a cohort of forward-looking AEC leaders committed to exchanging best practices and lessons learned in creating a culture of innovation in their company. This is an opportunity for face-to-face conversations and networking with like-minded peers. Learn more about the program at www.yegatech.com/aec-disruptors-circle/
- Keynotes: If you're looking for a keynote speaker for your annual business planning meetings or your next industry conference, we'd be happy to help you. We'll give your audience a balance of practical and inspirational messages.
- Disrupt It program: If you need help with forming your innovation team, creating a culture of innovation in your company, and creating innovation strategy and execution plans, we offer a fully custom program for your organization. For more details, visit www.yegatech.com
- Newsletter: Join our newsletter to stay informed about the latest status of AI, industry trends, and expert insights. You will have

access to exclusive content and receive updates on upcoming events. To subscribe, you can simply scan the QR code below:

Contact or follow us for the latest updates or ask any questions you may have at:

Dr. Sam Zolfagharian
- LinkedIn: www.linkedin.com/in/samzolfagharian/
- Website: www.samzolfagharian.com
- Twitter: @SZolfagharian

Dr. Mehdi Nourbakhsh
- LinkedIn: www.linkedin.com/in/mehdinour/
- Website: www.mehdinour.com
- Twitter: @MehdiNour_AEC

We look forward to connecting with you.

Thank you!

Dr. Sam Zolfagharian
Dr. Mehdi Nourbakhsh

ASKING FOR A FAVOR

Thanks again for reading. If you enjoyed reading this book, please consider sharing it with others who might find it inspiring or helpful. Your recommendation could make a difference and open up meaningful conversations. We'd also love to hear your thoughts, whether it's a review, a comment, or just a quick note about what resonated with you. Your feedback not only helps us grow but also supports the reading community in discovering new ideas and perspectives. Thank you for being part of this journey!

Stay inspired and keep sharing,

Sam & Mehdi

ENDNOTES

1. Grand View Research. Artificial Intelligence Market Size, Share & Growth Report 2030. *Grandviewresearch.com*. [Online] [Cited: September 15, 2024.] https://www.grandviewresearch.com/industry-analysis/artificial-intelligence-ai-market.

2. Cardillo, Anthony. How Many Companies Use AI? *explodingtopics.com*. [Online] August 21, 2024. [Cited: September 15, 2024.] https://explodingtopics.com/blog/companies-using-ai.

3. Blanco, Jose Luis, et al. Artificial intelligence: Construction technology's next frontier. *McKinsey.com*. [Online] April 4, 2018. [Cited: October 28, 2024.] https://www.mckinsey.com/capabilities/operations/our-insights/artificial-intelligence-construction-technologys-next-frontier.

4. Randolph, Marc. *That Will Never Work: The Birth of Netflix and the Amazing Life of an Idea*. s.l. : Little, Brown and Company, 2019. 0316530204.

5. PWC. Sizing the prize: PwC's Global Artificial Intelligence Study: Exploiting the AI Revolution. *www.pwc.com*. [Online] https://www.pwc.com/gx/en/issues/data-and-analytics/publications/artificial-intelligence-study.html#.

6. Georgieva, Kristalina. AI Will Transform the Global Economy. Let's Make Sure It Benefits Humanity. [Online] January 14, 2024.

[Cited: September 15, 2024.] https://www.imf.org/en/Blogs/
Articles/2024/01/14/ai-will-transform-the-global-economy-lets-
make-sure-it-benefits-humanity.

7. Neufeld, Dorothy. Long Waves: The History of Innovation Cycles.
Visual Capitalist. [Online] June 30, 2021. [Cited: September 22, 2024.]
https://www.visualcapitalist.com/the-history-of-innovation-cycles/.

8. *Computing Machinery and Intelligence.* Turing, Alan. 236, 10 1, 1950,
Mind, Vol. LIX, pp. 433–460.

9. Edwardo. SNARC. *History of AI.* [Online] April 04, 2019. [Cited:
November 18, 2024.] https://historyof.ai/snarc/.

10. Bernstein, Jeremy. A.I. *The New Yorker.* [Online] The New Yorker,
December 06, 1981. [Cited: November 18, 2024.] https://www.newyo-
rker.com/magazine/1981/12/14/a-i.

11. Gilesarchive, Martin. The GANfather: The man who's given
machines the gift of imagination. *MIT Technology Review.* [Online]
MIT Technology Review, February 21, 2018. [Cited: November 17,
2024.] https://www.technologyreview.com/2018/02/21/145289/the-
ganfather-the-man-whos-given-machines-the-gift-of-imagination/.

12. Hu, Krystal. ChatGPT sets record for fastest-growing user
base - analyst note. *www.reuters.com.* [Online] Reuters, February
2, 2023. [Cited: September 22, 2024.] https://www.reuters.com/
technology/chatgpt-sets-record-fastest-growing-user-base-analyst-
note-2023-02-01/.

13. albanyinstitute.org. GE Monitor-Top Refrigerator. *www.
albanyinstitute.org.* [Online] [Cited: November 15, 2024.] https://

www.albanyinstitute.org/online-exhibition/50-objects/section/
ge-monitor-top-refrigerator.

14. World Economic Forum. How venture capital is investing in AI
in these top five global economies. [Online] World Economic Forum,
May 24, 2024. [Cited: August 21, 2024.] https://www.weforum.org/
agenda/2024/05/these-5-countries-are-leading-the-global-ai-race-
heres-how-theyre-doing-it/.

15. Microsoft Corporate Blogs. Microsoft and OpenAI extend
partnership. [Online] Microsoft, January 23, 2023. [Cited: September 21, 2024.] https://blogs.microsoft.com/blog/2023/01/23/
microsoftandopenaiextendpartnership/.

16. Kindig, Beth. Microsoft –AI Will Help Drive $100 Billion In
Revenue By 2027. *Forbes.com*. [Online] June 15, 2023. [Cited: September 21, 2024.] https://www.forbes.com/sites/bethkindig/2023/
06/15/microsoft-ai-will-help-drive-100-billion-in-revenue-by-2027/.

17. Campbell, Morgan. Strategic innovation and why it's important.
www.planview.com/. [Online] Plan View. [Cited: October 28, 2024.]
https://www.planview.com/resources/guide/strategic-planning-
deliver-value/strategic-innovation-and-why-its-important/.

18. Bean, Randy. How Moderna Is Embracing Data & AI To Transform Drug Discovery. *Forbes.com*. [Online] 3 25, 2024. [Cited: September 15, 2024.] https://www.forbes.com/sites/randybean/2024/
03/25/how-moderna-is-embracing-data--ai-to-transform-drug-
discovery/.

19. MIT Technology Review. I Was There When: AI helped create
a vaccine. [Online] August 26, 2022. [Cited: September 21, 2024.]

https://www.technologyreview.com/2022/08/26/1058743/i-was-there-when-ai-helped-create-a-vaccine-covid-moderna-mrna/.

20. D^3 Faculty. AI puts Moderna within striking distance of beating COVID-19. *Digital Data Design Institute at Harvard.* [Online] [Cited: September 21, 2024.] https://d3.harvard.edu/ai-puts-moderna-within-striking-distance-of-beating-covid-19/.

21. Moderna, Inc. Moderna Reports Fourth Quarter and Fiscal Year 2021 Financial Results and Provides Business Updates. [Online] February 24, 2022. [Cited: September 22, 2024.] https://investors.modernatx.com/news/news-details/2022/Moderna-Reports-Fourth-Quarter-and-Fiscal-Year-2021-Financial-Results-and-Provides-Business-Updates/default.aspx.

22. Kotter, John P. How do you create a culture of innovation? *Fastcompany.com.* [Online] April 3, 2012. [Cited: September 22, 2024.] https://www.fastcompany.com/3011931/john-p-kotter-how-do-you-create-a-culture-of-innovation.

23. O'Connor, Gina. Real Innovation Requires More Than an R&D Budget. *HBR.com.* [Online] Harward Business Review, December 19, 2019. [Cited: September 22, 2024.] https://hbr.org/2019/12/real-innovation-requires-more-than-an-rd-budget.

24. Blanco, Jose Luis, et al. Accelerating Growth in Construction Technology. [Online] McKinsey, May 3, 2023. [Cited: September 22, 2024.] https://www.mckinsey.com/industries/private-capital/our-insights/from-start-up-to-scale-up-accelerating-growth-in-construction-technology.

25. Christensen, Clayton. *The Innovator's Dilemma: When New Technologies Cause Great Firms to Fail.* s.l. : Harvard Business Review Press, 1997.

26. Serani, Deborah. If It Bleeds, It Leads: Understanding Fear-Based Media. *Psychology Today.* [Online] June 7, 2011. [Cited: September 22, 2024.] https://www.psychologytoday.com/us/blog/two-takes-depression/201106/if-it-bleeds-it-leads-understanding-fear-based-media.

27. 3M. 3M's 15% Culture. [Online] 3M. [Cited: September 22, 2024.] https://www.3m.co.uk/3M/en_GB/careers/culture/15-percent-culture/.

28. Sakkab, Larry Huston and Nabil Y. Connect and Develop: Inside Procter & Gamble's New Model for Innovation. *hbr.com.* [Online] Harvard Business Review, March 2006. [Cited: September 22, 2024.] https://hbr.org/2006/03/connect-and-develop-inside-procter-gambles-new-model-for-innovation.

29. Roy, Puja. How Many Times Thomas Alva Edison Failed while inventing the light bulb? *Vedantu.* [Online] November 22, 2022. [Cited: October 22, 2024.] https://www.vedantu.com/blog/how-many-times-edison-failed-to-invent-bulb.

30. Dweck, Carol. *Mindset: The New Psychology of Success.* s.l. : Random House, 2006.

31. Ovans, Andrea. What Is a Business Model? *HBR.com.* [Online] Harvard Business Review, January 23, 2015. [Cited: September 22, 2024.] https://hbr.org/2015/01/what-is-a-business-model.

32. RGI. What is Radical Innovation? [Online] [Cited: October 22, 2024.] https://www.reallygoodinnovation.com/glossaries/radical-innovation.

33. IBM. Komatsu. *IBM*. [Online] IBM. [Cited: September 22, 2024.] https://www.ibm.com/case-studies/komatsu.

34. Miller, Ben. Nearly Half of U.S. Cities Have Fewer Than 1,000 Residents. *govtech.com*. [Online] Government Technology, December 3, 2018. [Cited: September 22, 2024.] https://www.govtech.com/data/nearly-half-of-us-cities-have-fewer-than-1000-residents.html.

35. Sherry, Ben. Even John Deere Is Getting in on the AI Boom. *Inc.com*. [Online] 3 26, 2024. [Cited: September 22, 2024.] https://www.inc.com/ben-sherry/even-john-deere-is-getting-in-on-ai-boom-heres-how.html.

36. Etiemble, Frederic. Hilti: Shift from a product to a service business model. *www.strategyzer.com*. [Online] 4 27, 2020. [Cited: 9 22, 2024.] https://www.strategyzer.com/library/lessons-from-hilti-on-what-it-takes-to-shift-from-a-product-to-a-service-business-model.

37. Bean, Randy. How Moderna Is Embracing Data & AI To Transform Drug Discovery. *Forbes.com*. [Online] 3 25, 2024. [Cited: 9 22, 2024.] https://www.forbes.com/sites/randybean/2024/03/25/how-moderna-is-embracing-data--ai-to-transform-drug-discovery/.

38. FinTech Global. AXA and AWS unite to launch risk management platform. *FinTech Global*. [Online] April 8, 2024. [Cited: September 22, 2024.] https://fintech.global/2024/04/08/axa-and-aws-unite-to-launch-risk-management-platform/.

39. The Mercedes-Benz Group. The first automobile 1885–1886. *The Mercedes-Benz Group.* [Online] [Cited: October 30, 2024.] https://group.mercedes-benz.com/company/tradition/company-history/1885-1886.html.

40. Ford Motors. The moving assembly line. *https://corporate.ford.com/.* [Online] [Cited: September 22, 2024.] https://corporate.ford.com/articles/history/moving-assembly-line.html.

41. Autodesk. 4 Common Project Closeout Issues in Construction. *govdesignhub.com.* [Online] 11 9, 2023. [Cited: September 22, 2024.] https://govdesignhub.com/2023/11/09/4-common-project-closeout-issues-in-construction/.

42. Learn Microsoft. Establish an AI Center of Excellence. *learn.microsoft.com* [Online] Microsoft, October 21, 2024. [Cited: October 28, 2024.] https://learn.microsoft.com/en-us/azure/cloud-adoption-framework/scenarios/ai/center-of-excellence.

43. Bosworth, Andrew. Building with AI across all of Meta. *ai.meta.com.* [Online] Meta, June 02, 2022. [Cited: October 28, 2024.] https://ai.meta.com/blog/building-with-ai-across-all-of-meta/.

44. Xu, C. How to be UBER successful at enterprise scale AI. *Digital Innovation and Transformation.* [Online] Harvard Business School, April 21, 2020. [Cited: October 28, 2024.] https://d3.harvard.edu/platform-digit/submission/how-to-be-uber-successful-at-enterprise-scale-ai/.

45. Cross, Amanda. 26 Employee Recognition Statistics You Need To Know In 2024. *nectarhr.com.* [Online] January 22, 2024. [Cited:

September 22, 2024.] https://nectarhr.com/blog/employee-recognition-statistics.

46. Dale, Edgar. *Audio-Visual Methods in Teaching.* New York : Dryden Press, 1946.

ACKNOWLEDGMENTS

Writing this book has been an incredible journey for both of us. This book is the result of countless hours of research, reflection, and collaboration, and it would not have been possible without the support and inspiration of many industry leaders and organizations.

First and foremost, we want to express our deepest gratitude to our clients. Your trust and partnership have not only fueled our professional work but have also provided the real-world insights that underpin the ideas in this book. We are thankful for the opportunity to work alongside you to navigate the challenges and opportunities AI brings to the AEC industry. You are the ones at the forefront, driving change and setting new standards in the industry. Your vision and willingness to innovate will shape the future, and we are proud to be part of that journey with you.

We are very appreciative of the members of our AEC Disruptors Circle. Your forward-thinking perspectives and willingness to challenge the status quo have influenced our approach to this book. The discussions and collaborations within this group have been invaluable in shaping the narrative of *Disrupt It*.

We are also grateful to all the thought leaders interviewed in the book. Your willingness to share your experiences, knowledge, and perspectives brought depth and authenticity to our work. Thanks for your openness and generosity in contributing to this project. We couldn't have done it without your invaluable help and support.

A special acknowledgment goes to our beta readers, Vince DiPofi, Scott Thompson, Cathy Myers, Josh McDowell, and Tim O'Rourke, who read the manuscript and provided invaluable feedback on how to improve it. Without your thoughtful critiques and encouragement, we could not have improved this book as much as we did.

We would like to express our gratitude to our incredible team at YegaTech for their hard work and support throughout the process of writing this book. We are also grateful to our friends who offered valuable insights and encouragement along the way.

And to you, our dearest readers! We are profoundly grateful for your interest in this book. We hope that the insights and stories shared within these pages inspire you to think differently about the AEC industry's future and embrace AI's transformative potential.

We would also like to acknowledge the influence of nature on this work. Much of this book was conceived and written during our time in Yosemite, California, where the breathtaking beauty of the wilderness fueled our creativity and provided the much-needed clarity we needed to write. The stunning landscapes of Yosemite reminded us of the importance of innovation that not only serves humanity but also respects and preserves the environment.

Finally, we want to thank our beloved families, whose support and understanding made this journey possible. Your love and encouragement have been our constant source of strength and inspiration.

As a side note, huge thanks to our dog, Goofy, for dragging us away from our desks for walks (when we needed it even more than him) while writing this book. You kept us from losing our minds and probably helped get this book finished!

With sincere appreciation,

Dr. Sam Zolfagharian, and
Dr. Mehdi Nourbakhsh

ABOUT THE AUTHORS

Dr. Sam Zolfagharian is the co-founder and President of YegaTech. Sam is passionate about the future of work in an AI-driven economy. With a focus on how technology is reshaping industries, she is dedicated to helping businesses navigate the evolving landscape of work and innovation. Sam has over twenty years of experience developing technologies for the design and construction industry, including eight years leading Autodesk's manufacturing and construction product platforms.

With a background in structural engineering, construction technology, and artificial intelligence, she has a unique perspective on the intersection of technology and the AEC industry. Sam is an AI keynote speaker, author, and co-creator of YegaTech's AI innovation framework, which is now being used to develop AI strategies and AI governance across the AEC industry.

Sam has led AI summits, training, and interactive keynotes with various large and small AEC companies and associations globally. She holds a Ph.D. in Design Computing and Construction Management from Georgia Tech, a master's in construction management, and a bachelor's in civil and structural engineering. She is also a published author with more than thirty research papers (with 500+ citations) in peer-reviewed journals and conferences.

Dr. Mehdi Nourbakhsh is an author, speaker, and CEO of YegaTech. As the author of Amazon bestselling book *Augment It: How Architecture, Engineering, and Construction Leaders Leverage Data and Artificial Intelligence to Build a Sustainable Future*, Mehdi co-created new frameworks for understanding and integrating AI technologies in AEC, establishing a solid foundation for innovation in the field.

With more than a decade of experience in the research and development of innovative AI solutions in the AEC and manufacturing industry at YegaTech, Autodesk, and Georgia Tech, Mehdi brings a unique perspective to this space. He has developed several AI solutions that are used by tens of thousands of AEC and manufacturing professionals every day and has eight U.S. patents on the use of artificial intelligence in these industries.

As a keynote speaker, he brings his unique insights and perspective to industry events. Mehdi is a firm advocate for strategic innovation, believing that companies should shift from simply reacting to AI disruptions to actively investing in processes that foster continual innovation and strategic growth.

Mehdi holds a Ph.D. in design computing from Georgia Tech, a master's in computer science and construction management, and a bachelor's in civil and structural engineering. He was recognized as one of the "40 Under 40" by the Georgia Institute of Technology in 2022 for his contributions to innovation and entrepreneurship.

About YegaTech

YegaTech, a technology consulting firm, specializes in developing AI strategy, governance, and execution plans, and setting new business models for the AEC industry, helping companies adapt to the era of disruption.

We offer customized programs to equip AEC professionals with essential AI skills, which increases the capacity to handle complex projects, stay ahead of the curve, and achieve cost-efficiency in a rapidly changing landscape.

With our decades of experience working in the AEC industry and AI solution development, we have created a proven framework to operationalize AI in your company. By forming an AI task force, finding AI opportunities that give your company market differentiation, creating an AI strategy and execution plan, and defining an AI governance framework, we'll help you accelerate your business strategy, get ahead of your competitors, and lead the industry in the age of disruption.

Fun fact: 'Yega' is short for 'Yeganeh,' which means 'unique and novel' in Farsi. In that sense, YegaTech represents unique and novel technology. We believe companies should pursue innovation opportunities unique to them to differentiate themselves in the market.

www.ingramcontent.com/pod-product-compliance
Lightning Source LLC
LaVergne TN
LVHW092333060326
832902LV00008B/622